Neurorehabilitation: Looking Back and Moving Forward

Neurorehabilitation: Looking Back and Moving Forward

Editors

Grigorios Nasios
Lambros Messinis
Efthimios Dardiotis
Markos Sgantzos

MDPI • Basel • Beijing • Wuhan • Barcelona • Belgrade • Manchester • Tokyo • Cluj • Tianjin

Editors

Grigorios Nasios
Speech and Language Therapy
University of Ioannina
Ioannina
Greece

Lambros Messinis
Department of Psychology
Aristotle University of Thessaloniki
Thessaloniki
Greece

Efthimios Dardiotis
Department of Neurology
University of Thesally
Larissa
Greece

Markos Sgantzos
Department of Anatomy
University of Thesally
Larissa
Greece

Editorial Office
MDPI
St. Alban-Anlage 66
4052 Basel, Switzerland

This is a reprint of articles from the Special Issue published online in the open access journal *Healthcare* (ISSN 2227-9032) (available at: www.mdpi.com/journal/healthcare/special_issues/neurorehabilitation_healthcare).

For citation purposes, cite each article independently as indicated on the article page online and as indicated below:

LastName, A.A.; LastName, B.B.; LastName, C.C. Article Title. *Journal Name* **Year**, *Volume Number*, Page Range.

ISBN 978-3-0365-7861-3 (Hbk)
ISBN 978-3-0365-7860-6 (PDF)

Contents

About the Editors

Grigorios Nasios

Grigorios Nasios, MD, PhD, is a neurologist and clinical neurophysiologist. He graduated from Ioannina University's Medical School, Greece, in 1990. He completed his specialty training in clinical neurology at the University Hospital of Ioannina in 1996, and he received his doctoral degree in medicine at the same university in 2001. He worked as a postdoctoral research fellow at the Department of Neurology at Freiburg and Ulm Universities in Germany. His current academic position is associate professor at the Department of Speech and Language Therapy, University of Ioannina, Greece. He has more than 20 years of clinical and academic experience in treating people living with neurological diseases and teaching at various academic settings (undergraduate and postgraduate students, in seminars, as an invited speaker, and as a doctoral students' supervisor). He has published more than 100 research articles in reputed journals and presented more than 80 presentations at international congresses.

Dr. Nasios is active in research in the fields of cognitive and behavioral neurology, aphasiology, disorders of communication in neurological diseases, dysphagia in stroke, neurocognitive disorders, and cognitive rehabilitation.

Lambros Messinis

Lambros Messinis is a clinical neuropsychologist and associate professor of neuropsychology in the Department of Psychology, Laboratory of Cognitive Neuroscience, at the Aristotle University of Thessaloniki. He graduated from the University of Johannesburg (p. RAU), South Africa, majoring in psychology, applied psychology, and human physiology. He then obtained a postgraduate master's degree with an emphasis in clinical health psychology from the same university. He further obtained two doctoral (PhD) degrees, one from the University of Johannesburg, Behavioural Medicine-Clinical Psychology Section, Unit of Psychophysiology, in 1997, and a second from the University of Patras Medical School, Neurology Department, specializing in neuropsychology and neuropsychological rehabilitation, in 2017. He also completed post-doctoral research/fellowship between 1997 and 1999 in the Medical Physics Laboratory at the University of Patras Medical School, with an emphasis in neuropsychology and neuroimaging, and between 2018 and 2019 in the Neurology Department at the University of Ioannina Medical School, with an emphasis in neuropsychology and neuropsychological rehabilitation in multiple sclerosis patients. From 2003 until April 2021, he was employed at the University Hospital of Patras, in Greece, where he directed the Memory-Neuropsychology unit of the Neurology Department and the Neuropsychology Laboratory of the Psychiatric Department in the same hospital. He has published 135 articles in national and international peer-reviewed journals with a high citation record and impact factor. He is an ad hoc peer reviewer in a large number of international journals with a high reputation and a member of the Editorial Boards of several prestigious journals. He has participated in several national and international research protocols and organized scientific meetings and conferences. He is also on the advisory committees and review groups for several grant applications.

Efthimios Dardiotis

Efthimios Dardiotis is associate professor, Department of Neurology, Faculty of Medicine, University of Thessaly, and Head of the Department of Neurology, General University Hospital of Larissa, Greece. He graduated with a BA degree from the Faculty of Medicine, Aristotle University of Thessaloniki, Thessaloniki, Greece, Medical Licence Number: 92691. He specialized in clinical neurology at the Department of Neurology, General University Hospital of Larissa, Greece. He obtained his PhD in neurology at the Faculty of Medicine, University of Thessaly, Greece. He is a post-doctoral fellow at the Cyprus Institute of Neurology and Genetics (CING), Cyprus, and has an MA in health administration from the Hellenic Open University, Greece. Furthermore, he is a lecturer in neurology at the University of Thessaly, Larissa. Has an MA in research methodology in biomedicine, biostatistics, and clinical bioinformatics from the University of Thessaly, Greece. Efthimios Dardiotis is an assistant professor at the Department of Neurology, Faculty of Medicine, University of Thessaly, and Head of the Department of Neurology, General University Hospital of Larissa, Greece. He is Head of the multi-disciplinary Multiple Sclerosis Center of the Department of Neurology, General University Hospital of Larissa, Greece.

Major scientific interests are: genetic epidemiology of multiple sclerosis (member of International Multiple Sclerosis Genetics Consortium, IMSGC); genetic epidemiology of amyatrophic lateral sclerosis (member of the NYGC ALS Consortium); epidemiology of Guillain-Barré Syndrome (national coordinator of International Guillain-Barré Syndrome Outcome Study, IGOS); epidemiology of Alzheimer's disease (member of the HELIAD study); genetic basis and epidemiology of Parkinson's disease; role of rTMS in cognitive rehabilitation (collaboration with Prof. Tsapkini, Department of Neurology, Johns Hopkins School of Medicine, Department of Cognitive Science, Baltimore, USA).

Markos Sgantzos

Markos Sgantzos is a Senior European Board Certified PRM (SEBPRM). He has a certificate of specialization in physical and rehabilitation medicine in Greece (since 1996). Delegate for the Section and Board of UEMS PRM and delegate for the European Society of Physical and Rehabilitation Medicine. He is vice president of the Hellenic Society of Physical and Rehabilitation Medicine; president of the examining committee for the medical speciality of Physical and Rehabilitation Medicine in Greece; member of the working group of KESY for the speciality of Physical and Rehabilitation Medicine (for the Ministry of Health); and consultant physiatrist, Department of Neurology, General University Hospital of Larissa. Associate Professor, Department of Anatomy, Faculty of Medicine, University of Thessaly Teaching "Physical Medicine and Rehabilitation" (4th year medical students), Faculty of Medicine, University of Thessaly (since 09/2010). He teaches anatomy and is a member of the faculty (1st and 2nd year medical students), Faculty of Medicine, University of Thessaly (since 12/2002). He is teaching in 10 postgraduate programs and is a member of the General Assembly of the Faculty of Medicine, University of Thessaly. He is a member of the committee of the accessibility center "PROSVASI" to enhance the physical, academic, and social access of students with disabilities at the University of Thessaly. Furthermore, he is responsible for the program for medical students with disabilities, a supervisor in six doctoral theses, a member of the advisory committee for 14 doctoral theses, a supervisor in 14 master's theses advisory committees (postgraduate study programs), and the author of four book chapters. He has performed over 47 lectures in national and international congresses; over 20 scientific and organizing congress committees (national and international congresses); editorial and advisory committees in two journals; twelve book translations into Greek; and over 131 publications.

Preface to "Neurorehabilitation: Looking Back and Moving Forward"

Neurorehabilitation is a complex, medically driven set of therapeutic-targeted actions applied to people who live with neurological diseases in order to restore and/or compensate for their disabilities. Progress in neurorehabilitation has been impressive over the years, but mainly in the last two decades. It follows the progress of clinical neuroscience in general, which is accelerated by a number of factors, with technology being a major one. Looking back, we are satisfied by the transition from experience-based, empirical neurorehabilitation to an era of evidence-based practice, in which multidisciplinary approaches and translational research are applied even from the acute phase through the disease course to an increasing number of people who need them worldwide. However, despite this progress, there are still many limitations and restrictions. Nervous system damage frequently poses difficult recovery issues, and research still has a lot to clarify about the system's neuroplasticity and reorganization dynamics. Additionally, more treatment approaches and combinations of therapeutic interventions need to be standardized, offered on a widespread basis, and individualized by a new generation of potent, skilled clinical neuroscientists, physicians, neuroradiologists, neuropsychologists, and other therapists. Furthermore, these interventions must be organized under a fresh "patient-centric" approach, with home-based and tele-rehabilitation protocols replacing long stays in medical centers. In other words, we have to move forward to a new era where medical, basic research, engineering, and high-tech advances can develop more efficacious approaches that can be widely applied to all people that need them.

In this Special Issue reprint, we present a collection of 11 empirical, clinical, basic research, and review papers covering the whole range of neurorehabilitation.

Grigorios Nasios, Lambros Messinis, Efthimios Dardiotis, and Markos Sgantzos
Editors

healthcare

MDPI

Editorial

Neurorehabilitation: Looking Back and Moving Forward, 1st Edition

Grigorios Nasios [1,*], Lambros Messinis [2], Efthimios Dardiotis [3] and Markos Sgantzos [4]

1 Department of Speech & Language Therapy, School of Health Sciences, University of Ioannina, 45500 Ioannina, Greece
2 Laboratory of Cognitive Neuroscience, School of Psychology, Aristotle University of Thessaloniki, 54124 Thessaloniki, Greece; lmessinis@psy.auth.gr
3 Department of Neurology, Faculty of Medicine, University of Thessaly, 41110 Larissa, Greece; ebsdar@gmail.com
4 Department of Anatomy, Faculty of Medicine, University of Thessaly, 41110 Larissa, Greece; sgantzosmarkos@gmail.com
* Correspondence: grigoriosnasios@gmail.com

Citation: Nasios, G.; Messinis, L.; Dardiotis, E.; Sgantzos, M. Neurorehabilitation: Looking Back and Moving Forward, 1st Edition. *Healthcare* **2023**, *11*, 1452. https://doi.org/10.3390/healthcare11101452

Received: 30 April 2023
Accepted: 15 May 2023
Published: 17 May 2023

Rehabilitation is "a set of interventions designed to optimize functioning and reduce disability in individuals with health conditions in interaction with their environment", according to the recent report from the World Health Organization (WHO), released in January 2023 [1]. As was pointed out in this report, rehabilitation is not "a luxury health service", but rather an "investment, with cost benefits for both the individuals and society", and potentially all of us need it, or will need it. As it is estimated, globally, about 2.4 billion people are currently living with a health condition that may benefit from rehabilitation [1], many of them with neurological diseases, to whom neurorehabilitation services should be offered. As was stated in our proposal, when releasing this Special Issue, progress in neurorehabilitation has been impressive over the years, but mainly in the last two decades. It follows the progress of clinical neurosciences in general, which is accelerated by several factors, technology being a major factor. Moving forward from empirical approaches, to an era of evidence-based practice, in which multidisciplinary approaches and translational research are applied even from the acute phase, through the disease course, numerous institutes and research groups have attempted to illuminate how nervous system plasticity and reorganization dynamics can be manipulated to offer better treatments, mainly non-pharmacological and neurobehavioral treatments. The lack of pharmaceutical remedies for the management of such disorders and the importance of cognitive or neuropsychological rehabilitation, neuromodulation, and multidisciplinary approaches were the focus of our attention in another recent Special Issue [2]. This fast-changing landscape of neurological rehabilitation, posing the need for a close follow up of the literature, and updated guidelines, was underlined via numerous examples and three of them are briefly reported below.

In 2019, we published a review article entitled "From Broca and Wernicke to the Neuromodulation Era... ", focusing on "methods for coupling new knowledge regarding the functional reorganization of the brain with sophisticated techniques capable of activating the available supportive networks in order to provide improved neurorehabilitation strategies" for recovery from aphasia [3]. Since then, low frequency repetitive transcranial stimulation (LF-rTMS) of the right inferior frontal gyrus (IFG) holds level B evidence for use in chronic poststroke non-fluent aphasia [4]. Furthermore, many investigators worldwide have published and currently are running protocols applying non-invasive transcranial magnetic or electrical stimulation, under different scenarios, regarding the brain region(s) targeted, the nature of lesions (vascular or neurodegenerative), and the time after aphasia onset.

A second example is rehabilitation of cognitive impairment in people living with multiple sclerosis (pwMS). While "no evidence" to support cognitive rehabilitation was the conclusion of a review of the literature in 2012 [5], there was a turn to "some evidence" in 2016 [6], with the "U-turn" completed in 2021, when the same authors changed their recommendation to "there is evidence" to support the rehabilitation of memory in pwMS [7]. It must be noted here that within "only" ten years of research progress, clinical practice was revolutionized.

A third example is neglect; even though it is a frequent consequence of right hemispheric damage [8], with a heavy impact on prognosis, and despite the existing evidence supporting its rehabilitation [4], it seems that many of us "neglect" it, being highly underdiagnosed and untreated [9]. In this case, it appears that the research has failed to "inform" clinical practice, at least not widely.

In this Special Issue of *Healthcare*, we present eleven novel articles from international groups of colleagues, including five clinical studies, five review articles, and one brief report.

Mariacristina Siotto et al. studied gender differences in 61 subacute stroke patients in terms of oxidative stress status, and the possible correlation between biomarkers of oxidative stress status and motor impairment, disability, and pain [10]. Measurements of hydroperoxide levels (d-ROMS), antioxidant activity (BAP test) and relative antioxidant capacity (OSI index) glucose levels and the lipid profile of the patients were performed. High levels of d-ROMS were recorded in all the patients, independently of gender, while BAP levels were in the normal range. Statistically significant gender differences were noted in the BAP and OSI values, with women having lower values, while in the male group, there was a negative correlation between oxidative stress status and motor impairment. On the other hand, a completely different trend of correlation was observed in the female group, maybe due to unbalanced systemic oxidative stress. These gender differences in terms of oxidative stress status should trigger further research in post-stroke patients in rehabilitation settings.

The second article, hosted in this Special Issue, is an elegant research study by Wohno Choi, investigating the effectiveness of cognitive exercise therapy on upper extremity sensorimotor function and daily functioning in 30 patients with chronic stroke, randomly divided into an experimental group (n = 15) and control group (n = 15) [11]. The experimental group received a combination of cognitive exercise therapy and conventional occupational therapy, while the control group received conventional occupational therapy exclusively. The authors found that only the experimental group improved significantly in sensory function, motor function and activities of daily living, in contrast to the control group, which improved only in the domain of daily activity. They concluded that the combined application of cognitive exercise therapy and conventional occupational therapy seems to have stronger results in terms of improving sensory and motor functions in patients with chronic stroke.

Daniel Fernandez-Sanchis and his colleagues from Spain studied the cost-effectiveness of a single dry needling (DN) session in 23 patients with stroke in the chronic phase, divided into the intervention group (IG) and the sham group (SG) [12]. The two variables used for the cost-effectiveness study were the values of the EuroQol-5D questionnaire and the Modified Asworth Scale (MMAS) to obtain the percentage of treatment responders and the quality-adjusted life years (QALYs) for each alternative. The evaluation of costs was carried out from the prospective of the hospital, clinic, or health center. A favorable cost-effectiveness ratio of both the EUR/QALYs and EUR/responder for IG emerged from the data analysis. However, the sensitivity analysis did not confirm the dominance of dry needling (higher effectiveness with less cost) over sham dry needling. Through this study, the application of DN to the upper limb was highlighted as an affordable alternative method in patients with chronic stroke.

In a randomized controlled trial, Choi et al. investigated the clinical effectiveness of Backward or Forward Walking Training programs combined with motor tasks on the balance and gait function of children diagnosed with spastic hemiplegic cerebral palsy [13]. Twelve children were randomly assigned to the Forward Walking Training Group (FWT) (n = 6) and the Backward Walking Training Group (BWT) (n = 6). The exercise programs were conducted 3 times per week for 4 weeks with a duration of 40 min per session. After six weeks from the completion of the first intervention, crossover training was also conducted. There was a statistically significant improvement in the spatiotemporal gait parameters (velocity, step, and stride) in both groups. Additionally, statistically significant improvement was found in walking speed, and balance function; however, the improvement was more significant in the BWT group. The authors concluded the importance of including the BWT program to improve balance, and thus prevent fall injuries in this population.

Christos Bakirtzis and his colleagues from Greece performed a validation study evaluating the psychometric properties of the Greek version of the Multiple Sclerosis Work Difficulties Questionnaire-23 (MSWDQ-23) [14]. The study involved 196 patients with multiple sclerosis (MS), all full-time or part-time employees. In addition to completing the MSWDQ-23, participants also completed self-reported questionnaires on fatigue, mood, daily functioning, and quality of life. The study verified the three-factor structure of the Greek version of the MSWDQ-23. In addition, the convergent validity of the questionnaire was good, as greater difficulties in the work environment were associated with higher EDSS, lower performance in cognitive tests, more fatigue, anxiety, stress and depression, and worse quality of life. The internal consistency of the questionnaire was excellent. Through this study, the Greek version of the MSWDQ-23 was proven to be a valid tool, which can be used in the context of interventions aimed at improving the occupational status of people with multiple sclerosis.

In a review article, Polychronis et al. critically approached the relevant literature of drooling in patients with Parkinson's disease (PD), and the possible relationship between excessive drooling and other clinical manifestations of PD, such as cognitive impairment, sleep difficulties, autonomic dysfunction, constipation, and orthostatic hypotension [15]. The study concluded that excessive drooling in PD patients is due to a wide range of factors. In parallel, excessive drooling was linked with a decrease in DAT binding in the striatum following DaTSCAN imaging.

Another review by Alexandratou et al. investigated the long-term neuropsychological effects following temporal lobe epilepsy surgery (a mean/median > 5 years post-surgery follow-up), according to the results of eleven included studies, based on the inclusion criteria defined in [16]. Interestingly, although one of the immediate consequences of surgery was a decline in cognitive functions, at the long-term follow-up, most studies demonstrated cognitive stability. Successful control of seizures leads mainly to the recovery of cognitive functions rather than a continued decline. The potential role of more selective surgery procedures in limiting cognitive side effects after surgery was also highlighted in this review.

Aloizou et al. reviewed the available studies that have used repetitive transcranial magnetic stimulation (rTMS) for the treatment of the most common types of dementia [17]. It has been argued that the application of this method, either alone or in combination with pharmaceutical therapy and cognitive training, appears to produce positive results in terms of improving cognitive functions. Although most protocols mostly involve the dorsolateral prefrontal cortex (DLPFC), it is a matter of further investigation. At the same time, the application of rTMS seems to benefit patients with mild cognitive impairment (MCI). Through this review, it can be perceived that rTMS is a promising method to improve the clinical picture of dementia, especially in Alzheimer's disease. At the same time, this review will be a trigger for further research, which will investigate the potential effectiveness of rTMS not only in Alzheimer's disease, but also in other types of dementia (vascular dementia—VD, Lewy body dementia—LBD; frontotemporal dementia—FTD).

Paulina Magdalena Ostroska and her colleagues from Poland reviewed the telerehabilitation of post-stroke patients in the era of the COVID-19 pandemic [18]. Their review evaluated the effectiveness of telerehabilitation on the functional status of post-stroke patients according to the results of 10 studies published from 2019 to 2021. Based on the studies that have investigated the feasibility and effectiveness of telerehabilitation programs in post-stroke patients, it has been argued that teletherapy appears to be effective not only in terms of functional recovery, but also in terms of mental and cognitive status, thereby improving the quality of life of patients. Through this review, the positive effects of telerehabilitation have been highlighted, which appear to be comparable to those of hospital treatment. Teletherapy appears to be an alternative method to traditional treatment and its applicability is particularly important in times of constraints, such as those caused by the SARS-CoV-2 pandemic.

In an observational study, Poveda-Garcia et al. investigated the ability to visualize mental motor images and the relationship between mental imagery and motor deficits of the upper limbs after stroke [19]. The study involved 39 patients, in whom upper limb movement and function, cognitive functions, and the ability to visualize mental images and functionality in daily life were assessed. The ability to visualize mental motor images of the upper limbs was highly significantly correlated with movement, functionality, and strength. In addition, visuospatial skills showed a correlation with the ability to visualize mental images, as well as with upper limb movement. Through this study, the particularly important role of the development of mental motor imagery in rehabilitation contexts is highlighted.

Finally, Agost-Gozalez et al. conducted a systematic review comparing the efficacy of percutaneous electrical stimulation (PTNS) and transcutaneous electrical stimulation (TTNS) procedures on the posterior tibial nerve in adults with idiopathic overactive bladder syndrome (OAB) who present with urinary incontinence (UI) [20]. A total of 19 studies published from 2015 to 2020 were included based on the inclusion criteria defined. The combined application of TTNS or PTNS procedures with other methods, such as pharmaceutical therapy or exercise, appears to be highly effective in reducing urinary incontinence (UI), and seems to contribute to a significant improvement in patients' perception of quality of life. Regarding the comparison of the two treatments, PTNS and TTNS, both procedures seem to be equally effective, but TTNS is mainly preferred, as it is more comfortable for the patient. Through this review, not only two globally recognized treatment methods were compared but also possible combinations of these treatments with other methods were highlighted, which may bring about the best results in terms of treating adult OAB individuals with UI.

As is apparent to the reader, the articles published in this Special Issue cover a wide range of neurorehabilitation modalities, although they only represent a small proportion of the whole scene. We trust that they will assist clinicians to perform evidence-based interventions and illuminate the necessity of multidisciplinary collaboration. Furthermore, they share an important motivational drive, that is, the inspiration to move forward to a new era, where medical, basic research, engineering and high technology advances can develop more efficacious approaches, combining neuromodulation techniques under sophisticated neuroimaging guidance. These can be widely and remotely applied to all people that need them, organized under a fresh "patient-centric" approach, with home-based and tele-rehabilitation protocols replacing long stays in medical centers.

Of course, there are many issues that still need to be addressed and fulfilled. A major issue is the priority that nations in the world must give to rehabilitation; in the 2030 initiative, released in 2017, the WHO emphasized the largely unmet need for rehabilitation, since most people still have no access to such services, although they need them [21]. The initiative highlights the importance of strengthening health systems to provide rehabilitation, marking a new strategic approach for the global rehabilitation community, with ageing populations, and an increase in the number of people living with chronic disease [21].

Another major issue is the convincing results of both basic and clinical research; as we can clarify the underlying mechanisms of brain dysfunction during disease, we will also be able to understand the mechanisms under which interventions such as cognitive training, exercise, music, or neuromodulation can change the brain [22]. This could be a key factor in conducting clinical trials that test innovative rehabilitation treatments and result in convincing evidence to translate research into clinical practice, since the lack of empirical evidence remains one of the main limitations of this translation [23].

Closing this editorial, we would like to inform the reader of the new series of articles already published in the second edition of this Special Issue, and hope to provide our readers with further empirical evidence and points for discussion, once this new Special Issue is complete.

Author Contributions: All authors (G.N., L.M., E.D. and M.S.) participated in the conceptualization, writing—original draft preparation, writing—review and editing. All authors have read and agreed to the published version of the manuscript.

Acknowledgments: We are grateful to all authors who contributed their valuable articles and expert reviewers for their significant contribution. We would also like to sincerely thank the journal for assisting in all aspects to make this Special Issue possible.

Conflicts of Interest: The authors report no conflict of interest or private agreements with companies concerning the manuscripts in this Special Issue.

References

1. Available online: https://www.who.int/news-room/fact-sheets/detail/rehabilitation (accessed on 25 April 2023).
2. Messinis, L.; Kosmidis, M.H.; Nasios, G.; Dardiotis, E.; Tsaousides, T. Cognitive Neurorehabilitation in Acquired Neurological Brain Injury. *Behav. Neurol.* **2019**, *2019*, 8241951. [CrossRef] [PubMed]
3. Nasios, G.; Dardiotis, E.; Messinis, L. From Broca and Wernicke to the Neuromodulation Era: Insights of Brain Language Networks for Neurorehabilitation. *Behav. Neurol.* **2019**, *2019*, 9894571. [CrossRef]
4. Lefaucheur, J.-P.; Aleman, A.; Baeken, C.; Benninger, D.H.; Brunelin, J.; Di Lazzaro, V.; Filipović, S.R.; Grefkes, C.; Hasan, A.; Hummel, F.C.; et al. Evidence-based guidelines on the therapeutic use of repetitive transcranial magnetic stimulation (rTMS): An update (2014–2018). *Clin. Neurophysiol.* **2020**, *131*, 474–528. [CrossRef]
5. das Nair, R.; Ferguson, H.; Stark, D.L.; Lincoln, N.B. Memory Rehabilitation for people with multiple sclerosis. *Cochrane Database Syst. Rev.* **2012**, *14*, CD008754. [CrossRef]
6. das Nair, R.; Martin, K.J.; Lincoln, N.B. Memory rehabilitation for people with multiple sclerosis. *Cochrane Database Syst. Rev.* **2016**, *10*, CD008754. [CrossRef] [PubMed]
7. Taylor, L.A.; Mhizha-Murira, J.R.; Smith, L.; Potter, K.J.; Wong, D.; Evangelou, N.; Lincoln, N.B.; das Nair, R. Memory rehabilitation for people with multiple sclerosis. *Cochrane Database Syst. Rev.* **2021**, *10*, CD008754. [CrossRef] [PubMed]
8. Parton, A.; Malhotra, P.; Husain, M. Hemispatial neglect. *J. Neurol. Neurosurg. Psychiatry* **2004**, *75*, 13–21.
9. Durfee, A.Z.; Hillis, A.E. Unilateral Spatial Neglect Recovery Poststroke. *Stroke* **2023**, *54*, 10–19. [CrossRef]
10. Siotto, M.; Germanotta, M.; Santoro, M.; Canali, R.; Pascali, S.; Insalaco, S.; Cipollini, V.; Papadopoulou, D.; Antonacci, E.; Aprile, I. Oxidative Stress Status in Post Stroke Patients: Sex Differences. *Healthcare* **2022**, *10*, 869. [CrossRef]
11. Choi, W. Effects of Cognitive Exercise Therapy on Upper Extremity Sensorimotor Function and Activities of Daily Living in Patients with Chronic Stroke: A Randomized Controlled Trial. *Healthcare* **2022**, *10*, 429. [CrossRef]
12. Fernández-Sanchis, D.; Brandín-de la Cruz, N.; Jiménez-Sánchez, C.; Gil-Calvo, M.; Herrero, P.; Calvo, S. Cost-Effectiveness of Upper Extremity Dry Needling in Chronic Stroke. *Healthcare* **2022**, *10*, 160. [CrossRef] [PubMed]
13. Choi, J.-Y.; Son, S.-M.; Park, S.-H. A Backward Walking Training Program to Improve Balance and Mobility in Children with Cerebral Palsy. *Healthcare* **2021**, *9*, 1191. [CrossRef] [PubMed]
14. Bakirtzis, C.; Artemiadis, A.; Nteli, E.; Boziki, M.K.; Karakasi, M.-V.; Honan, C.; Messinis, L.; Nasios, G.; Dardiotis, E.; Grigoriadis, N. A Greek Validation Study of the Multiple Sclerosis Work Difficulties Questionnaire-23. *Healthcare* **2021**, *9*, 897. [CrossRef] [PubMed]
15. Polychronis, S.; Nasios, G.; Dardiotis, E.; Messinis, L.; Pagano, G. Pathophysiology and Symptomatology of Drooling in Parkinson's Disease. *Healthcare* **2022**, *10*, 516. [CrossRef] [PubMed]
16. Alexandratou, I.; Patrikelis, P.; Messinis, L.; Alexoudi, A.; Verentzioti, A.; Stefanatou, M.; Nasios, G.; Panagiotopoulos, V.; Gatzonis, S. Long-Term Neuropsychological Outcomes Following Temporal Lobe Epilepsy Surgery: An Update of the Literature. *Healthcare* **2021**, *9*, 1156. [CrossRef]
17. Aloizou, A.-M.; Pateraki, G.; Anargyros, K.; Siokas, V.; Bakirtzis, C.; Sgantzos, M.; Messinis, L.; Nasios, G.; Peristeri, E.; Bogdanos, D.P.; et al. Repetitive Transcranial Magnetic Stimulation in the Treatment of Alzheimer's Disease and Other Dementias. *Healthcare* **2021**, *9*, 949. [CrossRef]
18. Ostrowska, P.M.; Sliwinski, M.; Studnicki, R.; Hansdorfer-Korzon, R. Telerehabilitation of Post-Stroke Patients as a Therapeutic Solution in the Era of the COVID-19 Pandemic. *Healthcare* **2021**, *9*, 654. [CrossRef]
19. Poveda-García, A.; Moret-Tatay, C.; Gómez-Martínez, M. The Association between Mental Motor Imagery and Real Movement in Stroke. *Healthcare* **2021**, *9*, 1568. [CrossRef]
20. Agost-González, A.; Escobio-Prieto, I.; Pareja-Leal, A.M.; Casuso-Holgado, M.J.; Blanco-Diaz, M.; Albornoz-Cabello, M. Percutaneous versus Transcutaneous Electrical Stimulation of the Posterior Tibial Nerve in Idiopathic Overactive Bladder Syndrome with Urinary Incontinence in Adults: A Systematic Review. *Healthcare* **2021**, *9*, 879. [CrossRef]
21. Available online: https://www.who.int/initiatives/rehabilitation-2030 (accessed on 25 April 2023).

22. Murphy, T.H.; Corbett, D. Plasticity during stroke recovery: From synapse to behaviour. *Nat. Rev. Neurosci.* **2009**, *10*, 861–872. [CrossRef]
23. Ploughman, M. A new era of multiple sclerosis rehabilitation: Lessons from stroke. *Lancet Neurol.* **2017**, *16*, 768–769. [CrossRef] [PubMed]

healthcare

MDPI

Article

Oxidative Stress Status in Post Stroke Patients: Sex Differences

Mariacristina Siotto [1,*], Marco Germanotta [1], Massimo Santoro [2], Raffaella Canali [1], Simona Pascali [1], Sabina Insalaco [1], Valeria Cipollini [1], Dionysia Papadopoulou [1], Erika Antonacci [1] and Irene Aprile [1]

1 IRCCS Fondazione Don Carlo Gnocchi ONLUS, 50143 Florence, Italy; mgermanotta@dongnocchi.it (M.G.); racanali@dongnocchi.it (R.C.); spascali@dongnocchi.it (S.P.); sinsalaco@dongnocchi.it (S.I.); vcipollini@dongnocchi.it (V.C.); dpapadopoulou@dongnocchi.it (D.P.); eantonacci@dongnocchi.it (E.A.); iaprile@dongnocchi.it (I.A.)

2 Energy and Sustainable Economic Development, Division of Health Protection Technologies ENEA-Italian National Agency for New Technologies, 00123 Rome, Italy; massimo.santoro@enea.it

* Correspondence: msiotto@dongnocchi.it; Tel.: +39-0633086552

Abstract: After a cerebral stroke insult, there is an overproduction of Reactive Oxygen Species (ROS), which overcome the antioxidant defenses, causing further tissues damage. The status of oxidative stress in stroke patients over time, particularly in those undergoing rehabilitation treatments, has been poorly investigated. We analyzed the oxidative stress status in 61 subacute stroke patients (33 females and 28 males) admitted to our rehabilitation center by measuring, in serum: hydroperoxides levels (d-ROMs), antioxidant activity (BAP test), and the relative antioxidant capacity (OSI index). We also analyzed patients for glucose levels and lipid profile. In addition, we analyzed the correlation between oxidative stress status biomarkers and motor deficits, disability, and pain. Almost all patients showed high or very high levels of d-ROMs, while BAP levels were apparently in the reference range of normality. Females had lower BAP values (females: 2478 ± 379; males: 2765 ± 590; $p = 0.034$) and lower OSI index (females: 5.7 ± 1.9; males: 6.8 ± 1.9; $p = 0.043$). Moreover, in the male group, the correlation with motor impairment and disability showed a worsened motor performance when oxidative stress is higher. Female group, on the other hand, had an unexpected different trend of correlation, probably due to an unbalanced systemic oxidative stress. Further research is needed to see if sex differences in oxidative stress status in subacute stroke patients persist after rehabilitation.

Keywords: oxidative stress; antioxidant defense; Oxidative Status Index (OSI); hydroperoxides; stroke; rehabilitation

Citation: Siotto, M.; Germanotta, M.; Santoro, M.; Canali, R.; Pascali, S.; Insalaco, S.; Cipollini, V.; Papadopoulou, D.; Antonacci, E.; Aprile, I. Oxidative Stress Status in Post Stroke Patients: Sex Differences. *Healthcare* **2022**, *10*, 869. https://doi.org/10.3390/healthcare10050869

Academic Editors: Grigorios Nasios, Lambros Messinis, Efthimios Dardiotis and Markos Sgantzos

Received: 6 April 2022
Accepted: 6 May 2022
Published: 9 May 2022

1. Introduction

Stroke is the principal cause of disability [1,2] and the second major cause of death worldwide, with a high burden on patients, their families, and health-care systems [3]. Patients after stroke have a very heterogeneous clinical spectrum, with variable and often partial recovery of motor function after a rehabilitation treatment [4]. Indeed, from 30 to 60% of patients present functional deficits of the paretic arm after a rehabilitation program, resulting in impaired activities of daily living [5].

During an ischemic insult, the brain–blood flow interruption causes multiple inflammatory immune responses and a general oxidative stress, which can damage the brain cells due to a secondary free radicals' formation and lipid peroxidation [6]. This means that the brain is especially exposed to oxidative stress.

Oxidative stress is defined as the imbalance between oxidant and antioxidant species, in favor of oxidants [7]. The most abundant free radicals are reactive oxygen species (ROS) and reactive nitrogen species (RNS), which are very highly reactive molecules, due to their unpaired electron(s) in their external shell. The source of free radicals can be endogenous (nicotinamide adenine dinucleotide phosphate, myeloperoxidase, lypoxigenase, angiotensin II, imbalance in essential metal homeostasis) or exogenous (pollution, alcohol, tobacco, heavy or transition metals, drugs, cooking radiation, etc.) [8,9]. Our antioxidant system counteracts the free radical toxicity and consists of endogenous antioxidants (enzymes: SOD, catalase, glutathione peroxidase; and non-enzymatic molecules: bilirubin, vitamin E, beta carotene, albumin, and uric acid) and exogenous antioxidants (Vitamin C, Vitamin E, phenolic antioxidants, oil lecithin, selenium, and zinc). Thus, oxidative stress status can be defined as the individual equilibrium between pro- and antioxidants [10]. This equilibrium varies with the general health status, specific disease condition, aging, or physical activity.

Very few studies examined oxidative stress status in survivors of stroke who are undergoing rehabilitation [11–13] respect to stroke patients in the acute phase [14–16]. Thus, the aims of this study were to examine (i) the oxidative stress status in subacute stroke patients admitted to our rehabilitation center and (ii) the relationship between the oxidative stress status and motor impairment, disability, and pain, dividing subjects by sex.

2. Materials and Methods

2.1. Sample

In this study, 61 patients with first stroke (33 females and 28 males), with a mean age of 68 $\pm$ 15 years, admitted to our rehabilitation department between 2019 and 2020 were consecutively enrolled.

The inclusion criteria were as follows: (i) first ischemic or hemorrhagic stroke, documented by magnetic resonance imaging (MRI) or computed tomography (CT); (ii) age between 55 and 85 years; (iii) time latency (within 6 months from stroke).

The exclusion criteria were as follows: (i) a previous stroke; (ii) behavioral and cognitive disorders and/or reduced compliance interfering with active therapy.

The study design was approved by the Ethical Committee of Don Carlo Gnocchi Foundation, Milan, Italy on 13 March 2019 (FDG_6_13/3/19). Written informed consent was obtained from all patients after a detailed explanation of the study's aims and rehabilitation protocols (clinical trials identifier: NCT04223180).

2.2. Biochemical Analyses

The blood samples of patients were collected in the early morning (7:30–9:00 a.m.) after an overnight fast to standardize the assessment of those biochemical variables that are affected by the circadian cycle and food intake. Sera samples were separated by centrifugation (3000 rpm, 10 min, and 4 °C). They were then divided into 0.5 mL aliquots and rapidly stored at −80 °C. Subjects' samples and reference samples were thawed just before the assay. All the analyses of the serum were performed in duplicate.

The colorimetric determination of Hydro-peroxides content (ROOH, principally) was assessed by d-ROMs test (Diacron, Grosseto, Italy) on an integrated analytical photometer (Free Carpe Diem, Diacron, Grosseto, Italy). This test measures the photometric variation between the byproduct of hydroperoxides with iron (RO· and ROO·, as for Fenton reaction) and a substituted aromatic amine (solubilized in a chromogenic mixture). The values are expressed in arbitrary units (UCARR), with 1 UCARR corresponding to 0.08 mg/100 mL of hydrogen peroxide [17]. Reference values are between 250–300 UCARR, while 301–320 UCARR is considered borderline range, 321–340 UCARR low level oxidative stress, 341–400 UCARR middle level of oxidative stress, 401–500 UCARR high level of oxidative stress, and >500 UCARR very high level of oxidative stress (Diacron, Grosseto, Italy).

In the blood, the defense against noxious attack of reactive species, such as free radicals, is guaranteed by the antioxidant barrier, which includes exogenous (ascorbate, tocopherols, carotenoids, bioflavonoids, etc.) or endogenous (proteins, bilirubin, uric acid, cholesterol, GSH, etc.) compounds. Each mentioned antioxidant compounds possessed its action power to oppose, depending on reduction-oxidation potential, to the oxidant action of ROS. Such power is associated to the ability of plasma barrier components to give reducing equivalent (electrons or hydrogen atoms) to reactive species, avoiding the abstraction of hydrogen atoms from biomolecules and the generation of dangerous radical chains. It allows to measure the chemically active antioxidant capacity (scavengers) of the plasma barrier. In particular, it includes antioxidants of both exogenous nature (ascorbic acid and tocopherols) and endogenous nature (uric acid, bilirubin and albumin) [18]. The state of global antioxidant defenses and/or the effectiveness of specific antioxidant treatments in plasma was measured by a BAP test (Diacron, Grosseto, Italy) measured on an integrated analytical photometer (Free Carpe Diem, Diacron, Grosseto, Italy). This test is based on the capacity of a colored solution of ferric ions (Fe^{3+}, R2 reagent) complexed to a chromogen to decolor when the ferric ions Fe^{3+} are reduced to ferrous ions (Fe^{2+}). This reduction is generated by an adequate reducing system, that is, antioxidant as is plasma. Reference values in μmol/L of antioxidants is >2200, 2200–2000 is a borderline status, 2000–1800 is considered slight deficiency status, 1800–1600 deficiency status, and 1600–1400 high deficiency status, while <1400 is a very high deficiency status condition (Diacron, Grosseto, Italy).

The antioxidative/oxidative stress ratio was also calculated using the ratio equation: BAP/d-ROMs. In accordance with literature, we named this ratio the OSI index, which is an index of potential antioxidant capacity [19]. The criterion value for the BAP/d-ROMs ratio was set at 7.3. Accordingly, a value lower than 7.3 was defined as an oxidized type and a higher or equal one as a reduced type. A higher BAP/d-ROMs was considered preferable. For an extensive and elucidative review on oxidative stress indexes for the diagnosis of health or disease in humans, see the work of Sanchez-Rodriguez [20].

The glucose levels and lipid profile included Glucose, Cholesterol, HDL Cholesterol and triglycerides analyses (Diacron, Grosseto, Italy). Glucose was measured by an oxidase/peroxidase system; reference value: 70–105 mg/dL [21]. Total cholesterol was measured by means of oxidation from a cholesteroxidase to cholest-4-en-3-one; normal values are <200 mg/dL, borderline values are 200–240 mg/dL, and high value are >240 mg/dL [21]. Triglycerides was measured by a peroxidase-coupled method; reference value 40–165 mg/dL [21]. Direct HDL Cholesterol was measured with a new method of elimination, in which after an oxidative reaction eliminating VLDL and LDL, the HDL portion was transformed into a quinone derivative read at 600 nm [22]. The reference values are differentiated by sex for this assay: for males, normal levels are >55 mg/dL (females: >656 mg/dL); for males, levels of 35–55 mg/dL indicate a moderate risk (females: 45–65 mg/dL); and for males, values of 35 mg/dL indicate a high risk (females: 45 mg/dL). Finally, we calculated the ratio total cholesterol and HDL cholesterol ([Cholesterol]/[HDL cholesterol]) (normal reference range: for male < 5, for female < 4.5).

2.3. Motor, Disability, and Pain Assessment

Patients were evaluated at admission in our rehabilitation center by means of: (i) the modified Barthel Index (BI), an ordinal scale used to measure performance in activities of daily living (ADL), ranging from 0 to 100, with lower scores indicating increased disability [23]; (ii) the Fugl-Meyer Assessment for upper extremity (FMA-UE) [24] to evaluate motor function; (iii) the upper-extremity subscale of the Motricity index (MI) [25] to evaluate limb strength; (iv) the Deambulation Index (DI), adapted form (eight-point scale) of the physical therapy part of the Patient Evaluation Conference System. The eight-point scale ranges from 0 (not assessed) to 7 [26]; (v) the Numerical Rating Scale (NRS), a unidimensional measure of pain intensity to diagnose and quantify pain in adults, in which a respondent selects a number from 0 (no pain) to 10 (extreme pain) that best reflects the intensity of their pain [27].

2.4. Statistical Analysis

Data were not normally distributed, according to the Shapiro–Wilk test, and therefore, non-parametric analysis was performed.

The values of d-ROMs and BAP of the sample group were compared using *t*-test with the identical test values measured in an Italian sample of 322 healthy volunteers (190 males and 132 females) subjected to a "health check" obtained from information on health status, physical measurements, and blood test [28].

To examine the relationship between biochemical data and demographic and clinical data, the Spearman rho correlation coefficients or the Mann–Whitney U test were used, as appropriate. This analysis was carried out in the whole sample, and for men and women, separately. Similarly, differences between men and women in biochemical, demographic, and clinical data were investigated using the Mann–Whitney U test or the chi-squared test.

For all the statistical analysis, a p value lower than 0.05 was deemed significant. Statistical analysis was performed using SPSS (IBM SPSS Statistics for Windows, Version 25.0. Armonk, NY, USA).

3. Results

3.1. Participants and Baseline Characteristics

For the study, 61 patients were enrolled and evaluated at admission to our rehabilitation center. Table 1 reports, in the whole group, and for females and males separately, the demographic and clinical characteristics of the sample.

Table 1. Baseline characteristics of the whole group (n = 61) and of females (n = 33) or males (n = 28). Data are reported in mean ± standard deviation or number and percentage (%).

Baseline Characteristics	Whole Group (n = 61)	Females (n = 33)	Males (n = 28)	p Value
Age (years)	68 ± 15	72 ± 13	65 ± 17	0.176
Index stroke type				
Ischemic	48 (78.7%)	29 (87.9%)	19 (67.9%)	0.057
Hemorrhagic	13 (21.3%)	4 (12.1%)	9 (32.1%)	
Affected side				
Right	24 (39.3%)	9 (27.3%)	15 (53.6%)	0.036 *
Left	37 (60.7%)	24 (72.7%)	13 (46.4%)	
Smoking	9 (14.8%)	5 (15.1%)	4 (14.3%)	0.9243
Comorbidities				
Hypertension	42 (68.9%)	19 (57.6%)	23 (82.1%)	0.039 *
Type 2 Diabetes	16 (19.7%)	6 (18.2%)	6 (35.7%)	0.7506
Dyslipidemia	10 (16.4%)	6 (18.2%)	4 (14.3%)	0.6821
Heart disease	23 (37.7%)	15 (45.5%)	8 (28.6%)	0.1752
Time from stroke onset (days)	110 ± 37	110 ± 40	110 ± 33	0.745
Numerical Rating Scale (pain)	3 ± 3	4 ± 3	3 ± 2	0.093
Motor Assessment				
Modified Barthel Index (0–100)	47.0 ± 20.4	46.2 ± 20.4	48.0 ± 20.7	0.592
Deambulation Index	2.0 ± 1.7	1.8 ± 1.7	2.1 ± 1.8	0.400
Motricity Index	39.2 ± 26.6	39.8 ± 26.8	38.2 ± 27.1	0.881
Fugl-Meyer Assessment	20.3 ± 17.0	23.4 ± 17.4	15.5 ± 15.7	0.222

* p value < 0.05.

Comparing the two groups, the only differences were in terms of affected side (females had a larger proportion of left hemiplegia, while in males right and left hemiplegia had similar percentages) and hypertension (with a higher percentage in males).

3.2. Oxidative Stress Biochemical Analyses

Oxidative stress serum analyses for the whole group, and divided between females and males, are reported in Table 2. A significative difference in BAP and OSI values were found between females and males, with antioxidant defense and OSI index lower in females (p = 0.034 and p = 0.043, respectively). On the contrary, no differences were found in terms of d-ROMs (p = 0.176).

From the comparison with a group of Italian Healthy controls [28], we found that post stroke patients (Table 2) had higher values of d-ROMs (healthy females: 364.70 ± 85.90 UCARR; healthy males: 312.0 ± 52.30 UCARR; Figure 1). The BAP test resulted also higher in post stroke patients (healthy females: 2035.74 ± 412.28 µmol/L, while healthy males: 1945.03 ± 406.64 µmol/L), but our female group had lower BAP values compared to males (Table 2).

Table 2. Oxidative stress biochemical analyses.

Biochemical Analyses	Whole Group (n = 61)	Female (n = 33)	Male (n = 28)	p Value
d-ROMs (UCARR)	448 ± 119	474 ± 144	417 ± 73	0.176
BAP (µmol/L)	2610 ± 504	2478 ± 379	2765 ± 590	0.034 *
OSI = BAP/d-ROMs ratio	6.2 ± 2.0	5.7 ± 1.9	6.8 ± 1.9	0.043 *

Data are reported as mean ± standard deviation; * p value < 0.05.

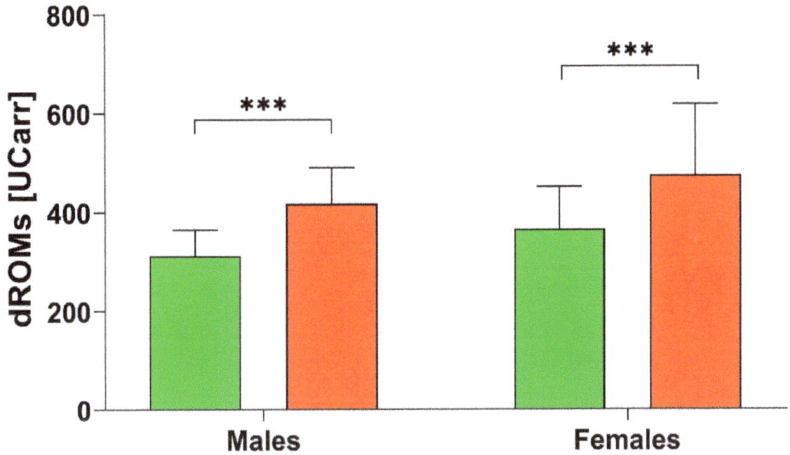

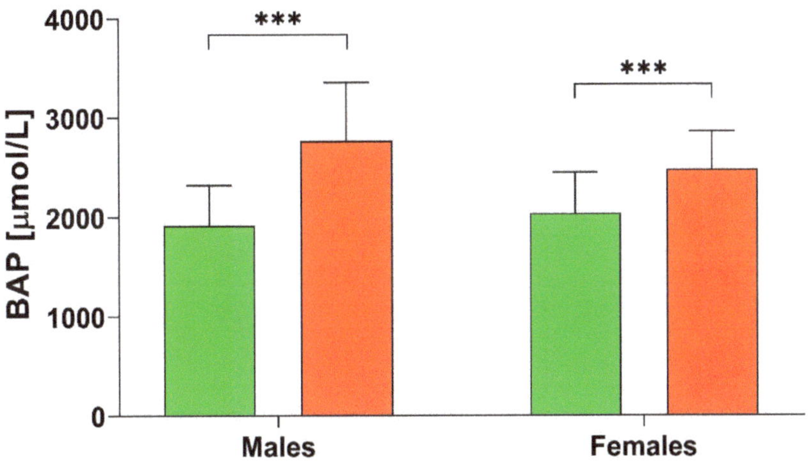

Figure 1. d-ROMs and BAP levels in healthy subjects [28] and post stroke patients divided by sex. *** $p < 0.001$ according to the t-test.

A descriptive analysis of the oxidative stress severity ranges, measured by d-ROMs serum test, are reported in Figure 2; a significative percentage of patients had high (28%) and very high (28%) percentages of oxidative stress severity. In particular, 43% of females had a very high oxidative stress level, while a similar percentage of males (43%) showed a high oxidative stress. The descriptive analysis of the antioxidant status ranges analyzed with BAP serum test are depicted in Figure 3, and revealed that most of the patients (79%) had an optimum power of counteracting oxidative stress (86% of males and 73% of females).

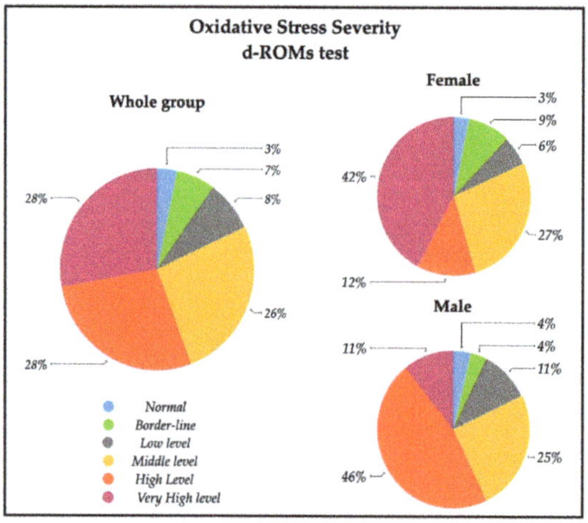

Figure 2. Pie charts representing the percentage of patients with oxidative stress severity in the whole group (n = 61), and in female patients (n = 33) and male patients (n = 28). Ranges are represented from normal to very high levels by means of d-ROMs test on serum. Normal (250–300 UCARR); Borderline (301–320 UCARR); Low level (321–340 UCARR); Middle level (341–400 UCARR); High level (401–500 UCARR); Very High level (>500 UCARR).

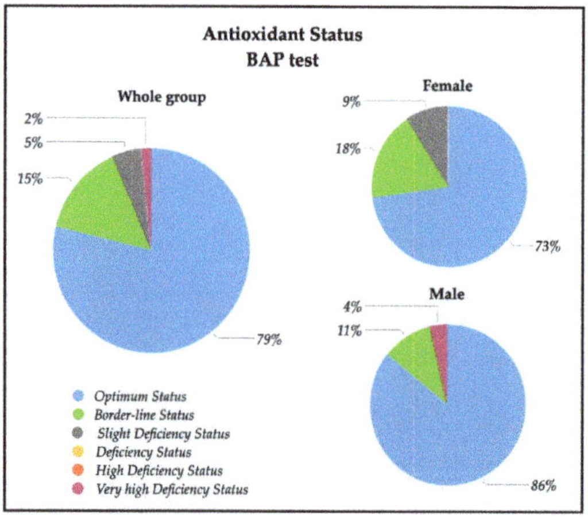

Figure 3. Pie charts of the percentage of patients with antioxidant status value in the whole group (n = 61), and in female patients (n = 33) and male patients (n = 28). Ranges are represented from optimum status to very high deficiency status measured by means of BAP test on serum. Optimum status (>2200 µmol/L); Borderline status (2200–2000 µmol/L); Slight deficiency status (2000–1800 µmol/L); Deficiency status (1800–1600 µmol/L); High Deficiency Status (1600–1400 µmol/L); Very high Deficiency Status (<1400 µmol/L).

The OSI index values was below the cut-off of 7.3 in 74% of patients (76% of females and 71% of males), indicating an insufficient antioxidant power with respect to hydroperoxide circulation.

3.3. Glucose Level and Lipid Profile Biochemical Analysis

Table 3 shows the biochemical analysis of the glucose levels and lipid profiles for the whole group, and divided by female and male. At the moment of admission in our structure, all of the patients had normal glucose, cholesterol, and triglyceride levels, with no sex differences. Only HDL cholesterol levels were different between males and females; while both were within normal reference ranges, males HDL cholesterol was lower and near the borderline value of 55 mg/dL, but the cholesterol ratio was normal.

Table 3. Glucose levels and lipid profile.

Biochemical Analyses	Whole Group (n = 61)	Female (n = 33)	Male (n = 28)	p Value
Glucose (mg/dL)	97.8 ± 42.2	95.1 ± 46.5	100.9 ± 37.2	0.473
Cholesterol (mg/dL)	119.0 ± 30.0	124.3 ± 28.3	113.0 ± 31.4	0.119
HDL Cholesterol (mg/dL)	64.4 ± 19.8	70.3 ± 19.7	57.5 ± 17.8	0.016 *
Cholesterol ratio (Cholesterol/HDL Cholesterol)	2.0 ± 0.7	1.9 ± 0.6	2.1 ± 0.7	0.300
Triglycerides (mg/dL)	113.7 ± 37.5	108.3 ± 36.4	119.9 ± 38.4	0.293

Data are reported as mean ± standard deviation; * p value < 0.05.

3.4. Correlation between d-ROMs, BAP, and OSI with Glucose Levels and Lipid Profile, Days from Index Stroke to Enrollment, and Motor, Disability, and Pain Assessment

The analysis of correlations between the hydroperoxides levels measured by means of d-ROMs levels is reported in Table 4. d-ROMs showed a positive correlation with glucose in the whole group and in female group. Moreover, a negative correlation with days from index stroke to enrollment, in the whole sample and in males. No correlations were found with the glucose levels and lipid profile, nor with pain, as measured by the NRS scale. In the whole group, a positive correlation was found between d-ROMs values and MI and FMA-UE; in females, d-ROMs values correlated positively with DI, MI, and FMA-UE; in males, a negative correlation was found between d-ROMs and BI and DI.

The analysis of correlations between antioxidant capacity by means of BAP levels (Table 5) showed in the male group a negative correlation with triglycerides and a positive correlation with the NRS. Correlation with motor assessment: in the whole group, BAP was found to be negatively correlated with the FMA-UE, while in the female group, a negative correlation was found with BI and DI.

Table 4. Correlation between the systemic hydroperoxides, measured by d-ROMs, and the glucose levels and lipid profile, days from index stroke to enrollment, pain scale, and motor and disability assessment in the whole group, and in female and male separately.

	d-ROMs					
	Whole Group (n = 61)		Female (n = 33)		Male (n = 28)	
	Spearman Rho	p Value	Spearman Rho	p Value	Spearman Rho	p Value
Glucose (mg/dL)	0.403 *	0.002	0.486 *	0.006	0.312	0.113
Cholesterol (mg/dL)	0.193	0.146	0.275	0.134	−0.059	0.770
HDL Cholesterol (mg/dL)	−0.054	0.689	−0.266	0.148	0.168	0.403
Triglycerides (mg/dL)	0.157	0.238	0.174	0.350	0.304	0.124
Time from Stroke onset (days)	−0.305 *	0.017	−0.258	0.147	−0.439 *	0.019

Table 4. *Cont.*

	d-ROMs					
	Whole Group (n = 61)		Female (n = 33)		Male (n = 28)	
	Spearman Rho	p Value	Spearman Rho	p Value	Spearman Rho	p Value
Numerical Rating Scale (pain)	−0.055	0.705	−0.226	0.258	0.049	0.827
Motor/disability Assessment						
Barthel Index	−0.004	0.975	0.254	0.154	−0.485 *	0.009
Deambulation Index	0.112	0.393	0.493 **	0.004	−0.505 *	0.006
Motricity Index	0.351 *	0.021	0.458 *	0.019	−0.009	0.974
Fugl-Meyer Assessment	0.379 *	0.016	0.495 *	0.014	−0.054	0.843

* p value < 0.05; ** p value < 0.005.

Table 5. Correlation between the total antioxidant capacity, measured by BAP, and the glucose levels and lipid profile, days from index stroke to enrollment, pain scale, and cognitive and motor assessment in the whole group, in females, and in males.

	BAP					
	Whole Group (n = 61)		Females (n = 33)		Males (n = 28)	
	Spearman Rho	p Value	Spearman Rho	p Value	Spearman Rho	p Value
Glucose (mg/dL)	−0.040	0.763	0.137	0.463	−0.245	0.217
Cholesterol (mg/dL)	−0.012	0.930	−0.071	0.703	−0.115	0.570
HDL Cholesterol (mg/dL)	0.050	0.712	0.202	0.274	0.120	0.550
Triglycerides (mg/dL)	−0.247	0.062	0.073	0.695	−0.569 **	0.002
Time from stroke onset (days)	0.042	0.749	−0.088	0.627	0.167	0.395
Numerical Rating Scale (pain)	0.008	0.957	−0.066	0.745	0.436 *	0.043
Motor Assessment						
Barthel Index	−0.121	0.353	−0.416 *	0.016	0.076	0.700
Deambulation Index	−0.088	0.505	−0.391 *	0.027	0.125	0.527
Motricity Index	−0.111	0.480	−0.261	0.197	0.140	0.591
Fugl-Meyer Assessment	−0.317 *	0.046	−0.379	0.068	−0.252	0.347

* p value < 0.05; ** p value < 0.005.

The OSI index correlated negatively with glucose and triglycerides in the whole group and in the male group. Moreover, in the whole group, a negative correlation was found with FMA-UE; in the female group, negative correlations were found with BI, DI, and MI, while positive correlations were found with BI and DI in the male group (Table 6).

Table 6. Correlation between the OSI index and glucose levels and lipid profile, days from index stroke to enrollment, pain scale, and cognitive and motor assessment in the whole group, in females, and in males.

	OSI Index					
	Whole Group (n = 61)		Females (n = 33)		Males (n = 28)	
	Spearman Rho	p Value	Spearman Rho	p Value	Spearman Rho	p Value
Glucose (mg/dL)	−0.277 *	0.035	−0.237	0.198	−0.415*	0.031
Cholesterol (mg/dL)	−0.082	0.538	−0.242	0.189	0.200	0.318
HDL Cholesterol (mg/dL)	0.078	0.560	0.306	0.094	0.001	0.997
Triglycerides (mg/dL)	−0.303 *	0.021	−0.153	0.410	−0.609 **	0.001
Time from Stroke onset (days)	0.211	0.102	0.140	0.436	0.295	0.128
Numerical Rating Scale (pain)	0.072	0.624	0.188	0.347	0.167	0.458

Table 6. *Cont.*

	OSI Index					
	Whole Group (n = 61)		Females (n = 33)		Males (n = 28)	
	Spearman Rho	p Value	Spearman Rho	p Value	Spearman Rho	p Value
Motor Assessment						
Barthel Index	−0.015	0.912	−0.392 *	0.024	0.422 *	0.025
Deambulation Index	−0.090	0.493	−0.581 **	0.000	0.475 *	0.011
Motricity Index	−0.288	0.061	−0.458 *	0.019	0.128	0.624
Fugl-Meyer Assessment	−0.460 **	0.003	−0.494 *	0.014	−0.267	0.318

* p value < 0.05; ** p value < 0.005.

No differences were found between smokers and no-smokers in oxidative stress markers. A negative correlation between d-ROMs and time from stroke onset was seen in the whole group and in the male group, showing that the further the distance from insult the lower the hydroperoxides levels.

4. Discussion

The analysis of oxidative stress status in the subacute stroke patients enrolled for this study revealed that systemic hydroperoxides levels were altered, as expected: 56% of the subjects had high and very high d-ROMs serum content (Figure 2, Table 2).

After an ischemia or a hemorrhagic brain injury, there is a massive production of ROS—as revealed by hydroperoxides in circulation—which can produce multiple reactions of radicals damaging the cells; lipid content of membrane cells is particularly susceptible to ROS attacks, because lipid peroxidation involves the inactivation of membrane enzyme and the destruction of the structural protein [29,30]. Several studies showed that acute and subacute ischemic stroke patients had increased levels of oxidative stress [11,12,31] and clinical severity of stroke was demonstrated to be correlated with increased serum hydroperoxide concentrations, measured with the d-ROMs test [16]. Moreover, free radicals also prevent recovery, which makes them an important post stroke therapeutic target [32].

Patients were admitted to our center after a certain time since stroke insult and we found negative correlation between d-ROMs and time from stroke onset in the whole group and in the male group, showing that the further the distance from insult the lower the hydroperoxides levels, as described in the literature [33]. Moreover, we found unexpectedly that the antioxidant levels were optimum (Table 2). In the whole group under study, in fact, the BAP test revealed a good antioxidant capacity in 79% of subjects (Figure 3). This result appears to point to a "counter-balanced oxidative stress" status, in which patients tend to neutralize the oxidative stress generated by stroke insult with an apparently good biological antioxidant capacity.

To investigate the redox balance more thoroughly, we calculated the OSI index for each patient; this index showed that the relative antioxidant capacity was not so effective, since 71% of the whole group had OSI under the cut-off (Table 2). Thus, the oxidant insult it is probably not sufficiently opposed by endogenous antioxidant capacity of patients. This scenario, however, is potentially harmful, because the oxidative stress could worsen if ROS production is not promptly removed or reduced [29]. There was no difference between smokers and no-smokers in oxidative stress biomarkers, and smokers were a very small percentage of patients, so we can exclude that hydroperoxides increase is related to smoking. It is also worth noting that the high mean age of our sample (68 ± 15 years; Table 1) can be a cause of worse antioxidant capacity.

We then compared our data to those of 322 Italian controls [28] and we found that our group had higher d-ROMs and BAP measurements (Figure 1). This "comparison group" was well defined in terms of age and gender, with no differences in d-ROMs values for age, and females having higher d-ROMs values. Antioxidant defenses decreased with age in the BAP test, with higher BAP values in females. The authors also divided females into premenopausal and postmenopausal groups but found no statistical differences [28].

Similarly, d-ROMs values were significantly different between 105 males and 185 females in another group of 290 "apparently healthy" Italians over 60, while PAT, which is an evolution of the BAP test, did not show significant differences [34].

In our patients, no differences were found in d-ROMs between males and females, but 42% of females had very high values of d-ROMs with respect to 11% of males (Figure 2). Females showed, instead, a lower level of BAP respect to males, with a lower percentage of normal levels of systemic antioxidants (Table 2, Figure 3). These findings show that oxidative stress status is different in the male and female group, but differently respect to healthy subjects. The higher systemic hydroperoxides levels measured with d-ROMs are most likely due to the stroke insult in both sexes, maintaining the same pattern of healthy subjects (with higher values in female), but our female group's antioxidant defense reservoirs appear lower to be lower.

Stroke incidence has long been known to be higher in males than females around the world; this sexually dimorphic epidemiology persists well past menopause until it is overshadowed by the effects of age [35,36]. Males have a higher age-adjusted stroke incidence than females [37], but females have not seen the same reduction in stroke rates as males, according to recent research [38]. Women's higher stroke rates may be due to longer life expectancies, but sex differences in stroke incidence rates may also play a role. A recent study on the X and Y chromosomes suggested that the second X chromosome has a negative effect that is only visible after reproductive senescence [39,40]. This suggests a complex interaction between aging, ischemia, and the sex chromosome, and that sex should be considered in the prevention and treatment of stroke [35,38,41].

The analysis of glucose levels and lipid profile, at the admission to our rehabilitation center, showed that all subjects had normal glucose, cholesterol, and triglyceride levels (Table 3), with no sex differences and no statistical differences between sex. Only HDL cholesterol levels differed between males and females being both within normal reference ranges; male HDL cholesterol was lower and near the borderline value of 55 mg/dL, but the total cholesterol/HDL cholesterol ratio remained within normal reference ranges. Brunelli et al. [42] analyzed sex differences in biomarkers in 195 Italian healthy volunteers, and found that males had lower HDL cholesterol values at the baseline, although the differences were not significant (males: 48.69 $\pm$ 14.10 mg/dL; female: 61.06 $\pm$ 11.78 mg/dL). In a more recent epidemiological study by Menotti [43], the analysis of 25,272 males (with median age of 51) and 21,895 females (median age 49) revealed higher levels in females.

The analysis of correlations with oxidative stress indices showed a positive correlation between d-ROMS and glucose in females and a negative correlation between BAP and triglycerides in males. These findings deserve further study to investigate better the relationship between glucose and lipid profile, as well as oxidative stress biomarkers, in subacute stroke male and female patients admitted to rehabilitation centers, considering that subjects after a stroke insult are often treated pharmacologically for hypercholesterolemia and diabetes. Note that among the comorbidities the hypertension is significantly more present in males than females and we do not able to exclude that this disease or the anti-hypertension drug assumption can have an influence on the oxidant stress status.

Another result of our study is that the analysis of correlation of d-ROMs and OSI index with motor assessment outcome showed very singular results: a substantial difference was observed between the female and male group. In the male group, the Barthel Index and Deambulation Index negatively correlated with d-ROMs and positively with the OSI index, showing that the higher the oxidative stress status was, the worse the motor function (Tables 3 and 5). Females had a totally different trend: the higher the d-ROMs values were, the better the ability to walk (measured using the Deambulation index) and the better the upper limb muscle strength performance (measured using the Motricity Index and Fugl-Meyer Assessment); the lower the antioxidant capacity was, the better the ability in daily activities and the better the deambulation (higher score in Barthel index and Deambulation index); the lower the OSI index was, which accounts for the antioxidant capacity index, the higher all the motor assessment outcomes were, i.e., Barthel index, Deambulation index, Motricity index, and the Fugl-Meyer Assessment (Tables 3–5). Males seem less distressed, and the oxidative stress status showed a significant correlation with motor ability. This particular behavior of the oxidative stress status in female is not associated with a higher disability or worse performance with respect to males because we found no significant differences in the relative outcomes. In our opinion, it is possible that females, after a certain level of oxidative stress, are no longer able to balance it, having an inadequate antioxidant capacity, and this implies that there is no correlation with motor assessment outcomes. This hypothesis has to be confirmed by measuring oxidative stress before and after rehabilitation, in order to clarify this point. In fact, the rehabilitation programs have been proven to be crucial in diminishing oxidative stress, lowering oxidative stress biomarkers in plasma [11,13,44,45]. In particular, a study on 29 subacute-phase post stroke patients reported levels of d-ROMs and BAP in plasma immediately before and after the exercises at admission and after a rehabilitation program, comparing an exercise group with a control one [12]. D-ROMs levels were significantly decreased and BAP levels significantly increased at rest only in the exercise group. It is interesting to notice that the levels of d-ROMs, measured immediately after the 1-h exercise, were always higher than before exercise. It is well known that high-intensity physical exercise is associated with increased production of reactive oxygen species, able to consume endogenous antioxidants and eventually able to damage biological molecules and key cellular components. Therefore, the balance between beneficial and potentially harmful effects of exercise might be of particular importance in the elderly, in which nutritional deficiencies, sedentary lifestyle, and comorbidity commonly concur with a depletion of the antioxidant reservoir of the organism and increased susceptibility to oxidative stress [46]. On the other hand, at rest, levels tend to diminish, showing the very positive effect of antioxidant activity exerted by exercise training. In this context, rehabilitation should be considered crucial not only for functional improvement and functional maintenance after stroke, but also for improving antioxidant capacity and attenuating systemic oxidative stress.

One limitation of this study is that we did not measure inflammation biomarkers or hormonal biomarker such as estrogen, which deserves future investigation for its correlation with oxidative stress. Another important limitation is that we tested only d-ROMs and BAP: the analysis of other biomarkers and an analysis on iron dysmetabolism need to be performed to extend and improve the study on oxidative status imbalance in post stroke patients. Furthermore, this study was carried out on a limited number of subjects and without an internal healthy control group; a study with a wider sample including a control group is planned to confirm these data. Moreover, we had no information about the cerebral area affected from the stroke, and this could be a critical point to evaluate the differences existing in oxidative stress among males and females.

From these preliminary data it is clear that further research is necessary to investigate the role of rehabilitation on oxidative stress status in post stroke patients and to evaluate if sex is a distinguishing factor of the response of the antioxidant ability to treatment. It is important to underline that the clinical picture after stroke can be heterogeneous and its evolution as well as the response to rehabilitation treatments can be very different in individuals, and in males and females, despite similar clinical status at the onset [47–49]. Reducing oxidative stress generated by free radicals is an important issue to consider in order to limit subsequent stroke insults. Moreover, further research is necessary to investigate the potential role of a specific diet or of antioxidant supplementation to overcome the oxidative imbalance reported in this study.

Author Contributions: Conceptualization, M.S. (Mariacristina Siotto), M.G., M.S. (Massimo Santoro) and I.A.; methodology, M.S. (Mariacristina Siotto); software, M.G.; validation, M.S. (Mariacristina Siotto), M.G. and M.S. (Massimo Santoro); formal analysis, M.S. (Mariacristina Siotto); investigation, M.S. (Mariacristina Siotto), M.G., D.P., S.I., S.P., E.A. and R.C.; data curation, M.S. (Mariacristina Siotto) and M.G.; writing—original draft preparation, M.S. (Mariacristina Siotto); writing—review and editing, M.S. (Mariacristina Siotto), M.G., M.S. (Massimo Santoro) and I.A.; visualization, R.C., S.P., V.C., S.I., D.P., E.A. and I.A.; supervision: I.A. All authors have read and agreed to the published version of the manuscript.

Funding: This research received no external funding.

Institutional Review Board Statement: The study was conducted according to the guidelines of the Declaration of Helsinki and approved by the Ethical Committee of Don Carlo Gnocchi Foundation, Milan, Italy on 13 March 2019 (FDG_6_13/3/19-clinical trials identifier: NCT04223180).

Informed Consent Statement: Written informed consent was obtained from all patients after a detailed explanation of the study's aims and re-habilitation protocols.

Data Availability Statement: The data supporting the findings of this study are available from the corresponding author upon reasonable request.

Conflicts of Interest: The authors declare no conflict of interest.

References

1. Kim, J.; Thayabaranathan, T.; Donnan, G.A.; Howard, G.; Howard, V.J.; Rothwell, P.M.; Feigin, V.; Norrving, B.; Owolabi, M.; Pandian, J.; et al. Global Stroke Statistics 2019. *Int. J. Stroke* **2020**, *15*, 819–838. [CrossRef] [PubMed]
2. Go, A.S.; Mozaffarian, D.; Roger, V.L.; Benjamin, E.J.; Berry, J.D.; Blaha, M.J.; Dai, S.; Ford, E.S.; Fox, C.S.; Franco, S.; et al. Executive Summary: Heart Disease and Stroke Statistics—2014 Update: A Report from the American Heart Association. *Circulation* **2014**, *129*, 399–410. [CrossRef] [PubMed]
3. GBD 2019 Stroke Collaborators Global, Regional, and National Burden of Stroke and Its Risk Factors, 1990–2019: A Systematic Analysis for the Global Burden of Disease Study 2019. *Lancet Neurol.* **2021**, *20*, 795–820. [CrossRef]
4. Prabhakaran, S.; Zarahn, E.; Riley, C.; Speizer, A.; Chong, J.Y.; Lazar, R.M.; Marshall, R.S.; Krakauer, J.W. Inter-Individual Variability in the Capacity for Motor Recovery after Ischemic Stroke. *Neurorehabilit. Neural Repair* **2008**, *22*, 64–71. [CrossRef]
5. Kwakkel, G.; Wagenaar, R.C.; Kollen, B.J.; Lankhorst, G.J. Predicting Disability in Stroke–a Critical Review of the Literature. *Age Ageing* **1996**, *25*, 479–489. [CrossRef]
6. Polidori, M.C.; Frei, B.; Cherubini, A.; Nelles, G.; Rordorf, G.; Keaney, J.F.; Schwamm, L.; Mecocci, P.; Koroshetz, W.J.; Beal, M.F. Increased Plasma Levels of Lipid Hydroperoxides in Patients with Ischemic Stroke. *Free Radic. Biol. Med.* **1998**, *25*, 561–567. [CrossRef]
7. Sies, H.; Berndt, C.; Jones, D.P. Oxidative Stress. *Annu. Rev. Biochem.* **2017**, *86*, 715–748. [CrossRef]
8. Liguori, I.; Russo, G.; Curcio, F.; Bulli, G.; Aran, L.; Della-Morte, D.; Gargiulo, G.; Testa, G.; Cacciatore, F.; Bonaduce, D.; et al. Oxidative Stress, Aging, and Diseases. *Clin. Interv. Aging* **2018**, *13*, 757–772. [CrossRef]
9. Siotto, M.; Filippi, M.M.; Simonelli, I.; Landi, D.; Ghazaryan, A.; Vollaro, S.; Ventriglia, M.; Pasqualetti, P.; Rongioletti, M.C.A.; Squitti, R.; et al. Oxidative Stress Related to Iron Metabolism in Relapsing Remitting Multiple Sclerosis Patients with Low Disability. *Front. Neurosci.* **2019**, *13*, 86. [CrossRef]
10. Sies, H. Oxidative Stress: Oxidants and Antioxidants. *Exp. Physiol.* **1997**, *82*, 291–295. [CrossRef]
11. Ciancarelli, I.; Pistarini, C.; Carolei, A.; Ciancarelli, M.G.T. Oxidative Stress in Post-Acute Ischemic Stroke Patients: Relevance of Early Intensive Neurorehabilitation. *J. Neurol. Neurophysiol.* **2013**, *9*, 266–273. [CrossRef]
12. Kihoin, N.; Tanaka, K.; Okuno, M.; Okamoto, T.; Saura, R. Exercise Attenuates Oxidative Stress in Patients with Stroke. *Neurol. Asia* **2016**, *21*, 7–16.
13. Hsieh, Y.W.; Lin, K.C.; Korivi, M.; Lee, T.H.; Wu, C.Y.; Wu, K.Y. The Reliability and Predictive Ability of a Biomarker of Oxidative DNA Damage on Functional Outcomes after Stroke Rehabilitation. *Int. J. Mol. Sci.* **2014**, *15*, 6504–6516. [CrossRef] [PubMed]
14. Koutsaliaris, I.K.; Moschonas, I.C.; Pechlivani, L.M.; Tsouka, A.N.; Tselepis, A.D. Inflammation, Oxidative Stress, Vascular Aging and Atherosclerotic Ischemic Stroke. Available online: https://pubmed.ncbi.nlm.nih.gov/34547993/#:~{}:text=Abstract, stiffness%2C%20among%20other%20pathophysiological%20features (accessed on 21 September 2021).
15. Wierońska, J.M.; Cieślik, P.; Kalinowski, L. Nitric Oxide-Dependent Pathways as Critical Factors in the Consequences and Recovery after Brain Ischemic Hypoxia. *Biomolecules* **2021**, *11*, 1097. [CrossRef]
16. Squitti, R.; Siotto, M.; Assenza, G.; Giannantoni, N.M.; Rongioletti, M.; Zappasodi, F.; Tecchio, F. Prognostic Value of Serum Copper for Post-Stroke Clinical Recovery: A Pilot Study. *Front. Neurol.* **2018**, *9*, 333. [CrossRef]
17. Cesarone, M.R.; Belcaro, G.; Carratelli, M.; Cornelli, U.; De Sanctis, M.T.; Incandela, L.; Barsotti, A.; Terranova, R.; Nicolaides, A. A Simple Test to Monitor Oxidative Stress. *Int. Angiol.* **1999**, *18*, 127–130.
18. Jansen, E.; Ruskovska, T. Serum Biomarkers of (Anti)Oxidant Status for Epidemiological Studies. *Int. J. Mol. Sci.* **2015**, *16*, 27378–27390. [CrossRef]
19. Imai, N.; Kobayashi, Y.; Uenishi, K. The Intake of Kiwifruits Improve the Potential Antioxidant Capacity in Male Middle- and Long-Distance Runners Routinely Exposed to Oxidative Stress in Japan. *Sports* **2021**, *9*, 37. [CrossRef]
20. Sánchez-Rodríguez, M.A.; Mendoza-Núñez, V.M. Oxidative Stress Indexes for Diagnosis of Health or Disease in Humans. *Oxidative Med. Cell. Longev.* **2019**, *2019*, 4128152. [CrossRef]

21. Trinder, P. Determination of Glucose InBlood Using Glucose Oxidase with an Alternative Oxygen Acceptor. *Ann. Clin. Biochem.* **1969**, *6*, 24–27. [CrossRef]
22. Izawa, S.; Okada, M.; Matsui, H.; Horita, Y. A New Direct Method for Measuring HDL-Cholesterol Which Does Not Produce Any Biased Values. *J. Med. Pharm. Sci.* **1997**, *37*, 1385–1388.
23. Shah, S.; Vanclay, F.; Cooper, B. Improving the Sensitivity of the Barthel Index for Stroke Rehabilitation. *J. Clin. Epidemiol.* **1989**, *42*, 703–709. [CrossRef]
24. Fugl-Meyer, A.R.; Jääskö, L.; Leyman, I.; Olsson, S.; Steglind, S. The Post-Stroke Hemiplegic Patient. 1. a Method for Evaluation of Physical Performance. *Scand. J. Rehabil. Med.* **1975**, *7*, 13–31. [PubMed]
25. Bohannon, R.W. Motricity Index Scores Are Valid Indicators of Paretic Upper Extremity Strength Following Stroke. *J. Phys. Ther. Sci.* **1999**, *11*, 59–61. [CrossRef]
26. Korner-Bitensky, N.; Mayo, N.; Cabot, R.; Becker, R.; Coopersmith, H. Motor and Functional Recovery after Stroke: Accuracy of Physical Therapists' Predictions. *Arch. Phys. Med. Rehabil.* **1989**, *70*, 95–99.
27. Downie, W.W.; Leatham, P.A.; Rhind, V.M.; Wright, V.; Branco, J.A.; Anderson, J.A. Studies with Pain Rating Scales. *Ann. Rheum. Dis.* **1978**, *37*, 378–381. [CrossRef]
28. Brunelli, E.; La Russa, D.; Pellegrino, D. Impaired Oxidative Status Is Strongly Associated with Cardiovascular Risk Factors. *Oxidative Med. Cell. Longev.* **2017**, *2017*, 6480145. [CrossRef]
29. Franzini, M.; Fornaciari, I.; Fierabracci, V. Indici di laboratorio di stress ossidativo (Rassegna). *LigandAssay* **2009**, *14*, 41.
30. Allen, C.L.; Bayraktutan, U. Oxidative Stress and Its Role in the Pathogenesis of Ischaemic Stroke. *Int. J. Stroke* **2009**, *4*, 461–470. [CrossRef]
31. Manolescu, B.N.; Berteanu, M.; Oprea, E.; Chiriac, N.; Dumitru, L.; Vladoiu, S.; Popa, O.; Ianas, O. Dynamic of Oxidative and Nitrosative Stress Markers during the Convalescent Period of Stroke Patients Undergoing Rehabilitation. *Ann. Clin. Biochem.* **2011**, *48*, 338–343. [CrossRef]
32. Miyamoto, N.; Maki, T.; Pham, L.-D.D.; Hayakawa, K.; Seo, J.H.; Mandeville, E.T.; Mandeville, J.B.; Kim, K.-W.; Lo, E.H.; Arai, K. Oxidative Stress Interferes with White Matter Renewal after Prolonged Cerebral Hypoperfusion in Mice. *Stroke* **2013**, *44*, 3516–3521. [CrossRef] [PubMed]
33. Žitňanová, I.; Šiarnik, P.; Kollár, B.; Chomová, M.; Pazderová, P.; Andrezálová, L.; Ježovičová, M.; Koňariková, K.; Laubertová, L.; Krivošíková, Z.; et al. Oxidative Stress Markers and Their Dynamic Changes in Patients after Acute Ischemic Stroke. *Oxidative Med. Cell. Longev.* **2016**, *2016*, e9761697. [CrossRef] [PubMed]
34. Gorni, D.; Finco, A. Oxidative Stress in Elderly Population: A Prevention Screening Study. *Aging Med.* **2020**, *3*, 205–213. [CrossRef] [PubMed]
35. Bushnell, C.; Howard, V.J.; Lisabeth, L.; Caso, V.; Gall, S.; Kleindorfer, D.; Chaturvedi, S.; Madsen, T.E.; Demel, S.L.; Lee, S.-J.; et al. Sex Differences in the Evaluation and Treatment of Acute Ischaemic Stroke. *Lancet Neurol.* **2018**, *17*, 641–650. [CrossRef]
36. Reeves, M.J.; Bushnell, C.D.; Howard, G.; Gargano, J.W.; Duncan, P.W.; Lynch, G.; Khatiwoda, A.; Lisabeth, L. Sex Differences in Stroke: Epidemiology, Clinical Presentation, Medical Care, and Outcomes. *Lancet Neurol.* **2008**, *7*, 915–926. [CrossRef]
37. Benjamin, E.J.; Virani, S.S.; Callaway, C.W.; Chamberlain, A.M.; Chang, A.R.; Cheng, S.; Chiuve, S.E.; Cushman, M.; Delling, F.N.; Deo, R.; et al. Heart Disease and Stroke Statistics-2018 Update: A Report From the American Heart Association. *Circulation* **2018**, *137*, e67–e492. [CrossRef]
38. Madsen, T.E.; Khoury, J.; Alwell, K.; Moomaw, C.J.; Rademacher, E.; Flaherty, M.L.; Woo, D.; Mackey, J.; De Los Rios La Rosa, F.; Martini, S.; et al. Sex-Specific Stroke Incidence over Time in the Greater Cincinnati/Northern Kentucky Stroke Study. *Neurology* **2017**, *89*, 990–996. [CrossRef]
39. Arnold, A.P.; Reue, K.; Eghbali, M.; Vilain, E.; Chen, X.; Ghahramani, N.; Itoh, Y.; Li, J.; Link, J.C.; Ngun, T.; et al. The Importance of Having Two X Chromosomes. *Philos. Trans. R. Soc. Lond. B Biol. Sci.* **2016**, *371*, 20150113. [CrossRef]
40. Banerjee, A.; Wang, J.; Bodhankar, S.; Vandenbark, A.A.; Murphy, S.J.; Offner, H. Phenotypic Changes in Immune Cell Subsets Reflect Increased Infarct Volume in Male vs. Female Mice. *Transl. Stroke Res.* **2013**, *4*, 554–563. [CrossRef]
41. Writing Group Members; Mozaffarian, D.; Benjamin, E.J.; Go, A.S.; Arnett, D.K.; Blaha, M.J.; Cushman, M.; Das, S.R.; de Ferranti, S.; Després, J.-P.; et al. Heart Disease and Stroke Statistics—2016 Update: A Report from the American Heart Association. *Circulation* **2016**, *133*, e38–e360. [CrossRef]
42. Brunelli, E.; Domanico, F.; La Russa, D.; Pellegrino, D. Sex Differences in Oxidative Stress Biomarkers. *Curr. Drug Targets* **2014**, *15*, 811–815. [CrossRef] [PubMed]
43. Menotti, A.; Puddu, P.E. Epidemiology of Heart Disease of Uncertain Etiology: A Population Study and Review of the Problem. *Medicina* **2019**, *55*, 687. [CrossRef] [PubMed]
44. Ciancarelli, I.; De Amicis, D.; Di Massimo, C.; Carolei, A.; Ciancarelli, M.G.T. Oxidative Stress in Post-Acute Ischemic Stroke Patients after Intensive Neurorehabilitation. *Curr. Neurovasc. Res.* **2012**, *9*, 266–273. [CrossRef] [PubMed]
45. Yeh, T.-T.; Hsieh, Y.-W.; Wu, C.-Y.; Wang, J.-S.; Lin, K.-C.; Chen, C.-L. A Preliminary Investigation of the Association of Sleep with Inflammation and Oxidative Stress Biomarkers and Functional Outcomes After Stroke Rehabilitation. *Sci. Rep.* **2017**, *7*, 8634. [CrossRef]
46. Polidori, M.C.; Mecocci, P.; Cherubini, A.; Senin, U. Physical Activity and Oxidative Stress during Aging. *Int. J. Sports Med.* **2000**, *21*, 154–157. [CrossRef]

47. Santoro, M.; Siotto, M.; Germanotta, M.; Bray, E.; Mastrorosa, A.; Galli, C.; Papadopoulou, D.; Aprile, I. BDNF Rs6265 Polymorphism and Its Methylation in Patients with Stroke Undergoing Rehabilitation. *Int. J. Mol. Sci.* **2020**, *21*, 8438. [CrossRef]
48. Santoro, M.; Siotto, M.; Germanotta, M.; Mastrorosa, A.; Papadopoulou, D.; Aprile, I. Association Study of SLC6A4 (5-HTTLPR) Polymorphism and Its Promoter Methylation with Rehabilitation Outcome in Patients with Subacute Stroke. *Genes* **2021**, *12*, 579. [CrossRef]
49. Siotto, M.; Germanotta, M.; Santoro, M.; Cipollini, V.; Guardati, G.; Papadopoulou, D.; Bray, E.; Mastrorosa, A.; Aprile, I. Serotonin Levels and Cognitive Recovery in Patients with Subacute Stroke after Rehabilitation Treatment. *Brain Sci.* **2021**, *11*, 642. [CrossRef]

healthcare MDPI

Review

Pathophysiology and Symptomatology of Drooling in Parkinson's Disease

Sotirios Polychronis [1,*], Grigorios Nasios [1], Efthimios Dardiotis [2], Lambros Messinis [3] and Gennaro Pagano [4,5]

1 Department of Speech & Language Therapy, School of Health Sciences, University of Ioannina, 451 10 Ioannina, Greece; grigoriosnasios@gmail.com
2 Department of Neurology, University Hospital of Larissa, University of Thessaly, 412 23 Larissa, Greece; ebsdar@gmail.com
3 Laboratory of Cognitive Neuroscience, School of Psychology, Aristotle University of Thessaloniki, 541 24 Thessaloniki, Greece; lmessinis@psy.auth.gr
4 Exeter Medical School, White City, London W12 0HR, UK; gennaro.pagano@roche.com
5 Neuroscience and Rare Diseases Discovery and Translational Area, Roche Pharma Research and Early Development, F. Hoffmann-La Roche Ltd., 4070 Basel, Switzerland
* Correspondence: polychronis.so@gmail.com; Tel.: +30-26510-50712 (ext. 50715); Fax: +30-26510-50732

Abstract: Drooling can present in patients with Parkinson's disease (PD), and it is manifested as an excessive pooling of saliva inside the oral cavity. Currently, the exact pathophysiological mechanism of drooling in PD is not yet fully explicated. Thus, it becomes crucial to understand if some clinical characteristics may emphasize drooling or if they are just concomitant. In PD, excessive drooling has been associated with a higher burden of non-motor symptoms, such as cognitive impairment, sleep problems, autonomic dysfunction, constipation and orthostatic hypotension, and of worse severity of motor fluctuations and bradykinesia. PD patients with excessive drooling also showed a reduction of striatal DAT availability at DaTSCAN imaging. Excessive drooling in patients with Parkinson's cannot be attributed to a single factor but to a mixture of factors, including but not limited to impaired nigrostriatal pathways.

Keywords: Parkinson's disease; salivation; clinical features; non-motor; motor; sialorrhea; drooling

Citation: Polychronis, S.; Nasios, G.; Dardiotis, E.; Messinis, L.; Pagano, G. Pathophysiology and Symptomatology of Drooling in Parkinson's Disease. *Healthcare* **2022**, *10*, 516. https://doi.org/10.3390/healthcare10030516

Academic Editor: José Carmelo Adsuar Sala

Received: 31 December 2021
Accepted: 8 March 2022
Published: 11 March 2022

1. Introduction

Drooling is commonly manifested among patients with Parkinson's disease, and it can be caused by the excess production of saliva, inability to retain saliva within the mouth (incontinence of saliva), or problems with swallowing (dysphagia) [1]. It can be caused due to a hyper-production of saliva inside the oral cavity or a change in salivary clearance as a result of swallowing impairments or difficulty in containing saliva inside the oral cavity [1]. Research studies have examined the pathophysiology of drooling in PD [2–16] in order to better understand the relationship between drooling and the clinical symptoms in PD. Over the course of the disease, the prevalence of drooling varied between 9.26% and 70% [17–21], it is higher in males [19,22–24] than in females and the longer the disease duration [17,18,24,25] and progression, the higher the risk of drooling [14,17,18,22,24–26]. Moreover, the higher the age and the severity of Levodopa-induced dyskinesia, the more prevalent the drooling is. A single work reported that drooling might, in some cases, be a prodromal PD symptom [20]. Even though inconsistent findings have been suggested regarding the association between the cognitive performance and drooling [17,22,23,25,27–30], it appears that drooling is related to sleeping disorders [22,23], dysautonomic symptoms [22,23,31], speech difficulties [23], dysphagia [17,23,25,32–35], hypomimia [24,26,36], bradykinesia [26], and a more symmetric pattern of PD presentation [24,37]. Neuroimaging research studies have suggested that de novo PD patients present with reduced functional connectivity in putamen, indicating that drooling is a symptom of a widespread pathology [38], which is challenging to treat [2]. Future research should further examine the relationship between drooling and other aspects of the PD symptomatology [39,40], as well as the influence of other treatments commonly used in PD and to analyze their consequences on drooling [41].

2. Pathophysiology of Drooling

The processes of salivation are controlled by both sympathetic and parasympathetic nervous systems [42]. The process of salivary gland secretion involves primarily cholinergic signaling by the parasympathetic nerves and signaling by neuropeptides, such as substance P, but also adrenergic signaling by sympathetic nerves. Parasympathetic stimulation will activate acetylcholine receptors, and sympathetic stimulation will increase alpha-receptor stimulation, which causes smooth muscle contraction and increases volume flow [42].

Drooling is more pronounced in periods where patients are "off" medication [2]. The abnormality of salivary production and maintenance inside the oral cavity and inadequate salivary clearance are considered the two main major domains that impact the pathophysiology of drooling [2]. Saliva's overproduction, by definition, can result in drooling. However, it has been found that PD patients produce less saliva compared to healthy controls [3–5], and the dopamine deficiency might explain why. However, the exact mechanism causing reduced salivary production has not been fully elucidated [4]. Studies using animal models have shown that saliva secretion is modulated by dopamine [6,7]. Specifically, the results of studies performed in rats have suggested that salivary secretion is a result of the activation of central and peripheral receptors of dopamine [7]. Lesion studies also support this claim, as a significant decrease in salivary secretion was identified when lesions in globus pallidus or its output pathway (lateral mesencephalic reticular formation) and the striatum were identified [8]. Aligned to this, a pathological study identified Lewy bodies in the superior cervical ganglion, the cervical sympathetic trunk, the peripheral vagus nerve, and the submandibular glands [9]. Another study by Costa et al. (2008) measured and compared the salivary production activity and velocity of salivary excretion of the parotic gland in healthy controls and PD patients, reporting the same production of saliva in both groups, but the parotic salivary excretion velocity to a distinct stimulus in the PD patients' group was significantly higher when compared to the group of the healthy controls [43]. Therefore, the increased velocity of saliva excretion should not be the main contributor of drooling in patients with PD, but it might partially contribute to its pathophysiology.

The other major domain that can contribute to drooling is swallowing dysfunction during the oral and/or pharyngeal phase of swallowing. In PD patients, bradykinesia can lead to oropharyngeal dysphagia. A study in animal models found a slower tongue protrusion in the rat group injected with 6-hydroxydopamine (6-OHDA) when compared to healthy controls [10]. In addition, the parkinsonian rat group in another 6-OHDA videofluorographic study had higher rates of aberrant food bolus movement compared to healthy controls [11]. Furthermore, a study using videofluoroscopy (VFSS) suggested that the severity of dysphagia in drooling PD patients is directly correlated with drooling [12]. Therefore, impairment in the oropharyngeal phage of swallowing might be a main contributor to drooling's pathophysiology. Furthermore, Kikuta T. et al. (2011) suggested that advanced PD patients present with lower maximum tongue pressure when compared to PD patients in early or moderate stages of their disease and that there is a negative correlation between the oropharyngeal transit time and the tongue movement [13]. Therefore, poor muscle control of the tongue and bradykinesia can contribute to the dysphagia's pathophysiology and possibly to drooling itself. Hypomimia, involuntary mouth opening, stooped posture of the upper body, and dropped head can affect the patients' ability to maintain saliva inside their oral cavity, and, therefore, cause drooling in PD [14]. Finally, studies using manometry suggested that impaired upper oesophageal sphincter (UES) mobility might also have an effect on dysphagia and drooling in PD patients. However, this cannot be the sole cause of dysphagia in patients who have adequate clearance mechanisms and pharyngeal propelling forces [15,16].

3. PD-Associated Symptomatology and Drooling

Drooling in PD patients correlates with other general clinical features, non-motor, and motor symptoms.

3.1. General Clinical Features

The prevalence of drooling ranges between 9.26% and 70% due to the disease heterogeneity and the different instrument measures used in every study [17–25,44–46]. It can manifest very early during the disease [20], but it is not considered a prodromal symptom in Parkinson's according to the current MDS research criteria. Based on Braak's staging of brain pathology in PD and the model of the hypothesized spread of alpha-synuclein (aSyn) in PD, aSyn accumulation begins in the gut and then progresses up to the brain via the vagus nerve [47]. Therefore, it can be suggested that gastrointestinal tract features should be a prominent early manifestation of PD. However, more scientific studies are required to better investigate whether drooling can contribute to the diagnosis of Parkinson's.

The prevalence of drooling is higher in males [19,22–24] than females since women with PD showed a less malignant phenotype [48], with the estrogen activity in females probably delaying the development of the PD symptoms [49].

Moreover, the longer the disease duration [17,18,24,25] and progression, the higher the risk of drooling [14,17,18,22,24–26]. As drooling is mainly a result of a decreased frequency of saliva clearance inside the oral cavity, problems with posture [5], and oral motors, as well as facial impairment [31] (e.g., bradykinesia, rigidity, hypomimia), it is considered that all these deficits are more common and severe as the disease progresses.

Age is another crucial factor when it comes to prevalence. Drooling becomes more prevalent with age [17,22,24,25]. Any age-related changes can impact saliva control as we grow older. With aging, a natural progressive brain tissue loss that links up with neurological skills worsening and muscle-mass reduction occurs [50]. Therefore, the decreased muscle strength of the orofacial muscles (e.g., buccinator, tongue, and orbicularis oris) [50] can result in the accumulation of saliva inside the oral cavity and increase the possibility of anterior and posterior spillage of saliva. However, given that these studies did not include a control group, more scientific studies are required for more reliable results.

The prevalence of drooling is higher in PD patients with higher severity of Levodopa-induced dyskinesia (LID) [22,23]. This is shown in more advanced patients, where generally higher levodopa doses are used [14,17,18,22,24–27,51,52].

3.2. Non-Motor, Motor Symptoms, and Drooling

It has been suggested that a variety of non-motor and motor symptoms can occur during the course of Parkinson's disease [53].

In regard to cognitive function, there have been inconsistent findings around the role of cognitive performance in drooling [17,22,23,25,28,29]; however, some studies suggested that drooling is linked to cognitive decline [27,30]. Specifically, Reynold et al. (2018) suggested that cognition has a role beyond the automatic process of drooling and saliva control [28]. They showed that divided attention impairment aggravates drooling in PD patients, using a paradigm where the vigilance of saliva control and frequency of saliva swallowing decreased during a distracting cognitive task [28]. Nevertheless, more studies are required to better understand if a closer link between cognition and drooling exists.

Sleeping disorders are associated with the appearance of drooling [22,23]. Good sleep quality in patients with PD was correlated with less motor symptoms in the morning [54]. Therefore, poor sleeping patterns can influence motors symptoms and promote drooling [22].

Dysautonomias, namely urinary disorders, sexual dysfunction [23], obstipation [22], other gastrointestinal manifestations, and orthostatic hypotension, were found to relate to drooling. The autonomic system is affected due to the alterations of the vagus nerve [31] and, therefore, it can result in several concomitant dysfunctions, such as drooling, gastrointestinal problems, and obstipation.

Speech difficulties [23] and dysphagia [17,23,25] are associated with drooling. Speech, swallowing, and saliva control share many organs and anatomical structures. Thus, impairment in one of them can result in dysfunction in all these areas. Muscles inside the oral cavity, as well as lips, tongue, jaw, cheeks, larynx, and pharynx, are impacted by rigidity, bradykinesia, and hypokinesia, usually present in PD [32,33]. A delayed swallowing reflux [34], lingual tremor, lingual pumping, prolonged lingual elevation, and mandibular excursion are observed changes in patients with PD and can contribute to reduced saliva control [35].

Hypomimia was also associated with drooling [24,26], and, therefore, lip closure reduction is noted in some of the PD patients affecting saliva control [36].

Bradykinesia was found to be linked to drooling [26], as it can impact the orofacial muscles [33]. As a result, the reduction of movements velocity of the lips, tongue, jaw, and cheeks, among others, can affect the control of saliva inside the oral cavity and its transport from the oral cavity to the oropharynx.

Patients with a PD dominant tremor did not present with a higher prevalence of drooling [22]. Nevertheless, according to one study, patients with PD non-dominant tremor presented with a higher prevalence of drooling [22]. These results can be attributed to the fact that non-dominant tremor patients exhibit a higher decrease in grey matter and neural functional connectivity related to motor regions [55], as well as an extensive Lewy bodies pathology in the cortical areas [56]. They also present with greater lingual control dysfunction and increased rigidity in the oropharynx [13].

Drooling was related to a more symmetric pattern PD presentation [24]. Patients with a higher burden of motor symptoms present with a more symmetric pattern [37], and, therefore, we can anticipate that drooling will also be more prevalent in such patients.

Interestingly, de novo PD patients with drooling have not been extensively investigated, and, therefore, drug treatment (e.g., Levodopa) might have affected the results. Levodopa is the main drug used in PD treatment; however, its long-term use can lead to dyskinesia and motor fluctuations [57]. Dyskinesia is generally progressive and can impact different areas of the body, including the orofacial muscles, the neck, the tongue, and the jaw [58]. Motor deficits resulting from dyskinesia in the aforementioned areas can enhance drooling.

In addition, the relation of DAT binding in the striatum has not been extensively investigated. Tajima et al. (2020) suggested that the severity of motor symptoms, especially axial symptoms (components of akinetic-rigid PD) and bradykinesia, but no tremor and specific binding ratio (SBR), in de novo PD may relate to drooling [59]. Therefore, it can be hypothesized that the mechanism of the drooling aggravation is similar to that of bradykinesia and axial symptoms since previous studies have shown that DAT binding correlates with bradykinesia and axial symptoms, but not with parkinsonian tremor [38,60,61]. However, the effects of Levodopa remain unclear [2,39], indicating that other mechanisms, in addition to the nigrostriatal dopamine system, play a role in drooling. An fMRI study (Hou et al., 2016) examined the functional connectivity in the basal ganglia of de novo PD patients with and without drooling. The drooling patients had significantly decreased functional connectivity in the putamen within sensorimotor cortices (bilaterally), the parietal (inferior and superior) lobules, and other areas in the occipital (right) and temporal (right) lobes [40]. Therefore, it can be inferred that drooling is a symptom of a widespread pathology, and it cannot be attributed to a single causing factor. As a result, the management of drooling is complex, as identifying treatment options to target such a widespread pathology can be challenging.

There have been suggested pharmacological and non-pharmacological treatments to tackle drooling in PD [2]. First, PD patients should withdraw cholinesterase inhibitors, namely quetiapine and clozapine, because they aggravate drooling [2]. Next, they should attempt to improve their motor symptoms (e.g., by using dopaminergic medication or performing deep brain stimulation (DBS)) [2]; nevertheless, it is important to highlight that there is no study that has specifically investigated the effect of DBS on PD patients with drooling. Behavioral modification and radiotherapy might also be used as complementary therapies [2]. However, all these treatments only partially contribute to the management of drooling and more specific treatment options are still required.

4. Limitations and Future Directions

It is eminent that future studies will use drooling-specific rating scales (e.g., Drooling Severity and Frequency Scale (DSFS), Sialorrhea Clinical Scale for PD (SCS-PD), and Drooling Rating Scale (DRS)) [2] as opposed to subjective data and patients' complaints to evaluate drooling in PD. They should also include the examination of biochemical properties of saliva, its appearance, viscosity, flow, and volume, as well as the association between the degree of drooling severity and patients' clinical characteristics. Interestingly, a study found that the Radboud Oral Motor inventory for Parkinson's disease—Saliva (ROMP-saliva) scale is the only scale with data on patients with PD and clinimetric properties adequacy [62].

Moreover, other important aspects need further research. It would be interesting to examine the association between saliva production and smell, given that saliva secretion rate can be affected by smell [41] since the smell of food usually increases saliva production [63], and hyposmia is commonly observed in patients with PD. In addition to this, the association drooling has with fatigue and sensory deficits (i.e., vision abnormalities), as PD patients usually complain about those during the course of their disease [53]. It is essential to better understand whether drooling affects early PD patients so that we assess whether a correlation with a worse PD phenotype at the later disease stages occurs. Furthermore, more specific guidance is required around the pharmacological treatment of drooling and how the administration of botulinum toxin as the standard pharmacological treatment in drooling can have positive or negative impacts on other clinical features. Thus far, it is known that anticholinergic drugs to reduce drooling can have side effects, such as hallucinations or delirium [64]. Finally, including a control group would allow for more reliable results and safer conclusions.

5. Conclusions

The exact pathophysiology of drooling in patients with PD has not yet been fully explicated. More knowledge regarding how drooling is associated with clinical characteristics will help us understand whether such factors worsen drooling or are just concomitant. Excessive drooling has been associated with greater burdens of non-motor symptoms and increased severity of motor fluctuations and bradykinesia. A decrease of DAT binding in the striatum at DaTSCAN imaging has also been shown. All in all, excessive drooling in patients with Parkinson's cannot be attributed to a single factor, but to a mixture of factors, as part of a widespread pathology that is complex to treat.

Author Contributions: Conceptualization, S.P., G.N. and G.P.; writing—original draft preparation, S.P.; writing—review and editing, S.P., G.N. and G.P.; supervision, G.N., G.P., E.D. and L.M. All authors have read and agreed to the published version of the manuscript.

Funding: This research received no external funding.

Institutional Review Board Statement: Not applicable.

Informed Consent Statement: Not applicable.

Conflicts of Interest: Pagano is a full-time employee of F. Hoffmann-La Roche Ltd. The other authors declare no conflict of interest.

References

1. Fasano, A.; Visanji, N.P.; Liu, L.W.C.; Lang, A.E.; Pfeiffer, R.F. Gastrointestinal dysfunction in Parkinson's disease. *Lancet Neurol.* **2015**, *14*, 625–639. [CrossRef]
2. Srivanitchapoom, P.; Pandey, S.; Hallett, M. Drooling in Parkinson's disease: A review. *Parkinsonism Relat. Disord.* **2014**, *20*, 1109–1118. [CrossRef]

3. Bateson, M.; Gibberd, F.B.; Wilson, R.S.E. Saliva symptoms in Parkinson disease. *Arch. Neurol.* **1973**, *29*, 274–275. [CrossRef]
4. Bagheri, H.; Damase-Michel, C.; Lapeyre-Mestre, M.; Cismondo, S.; O'Connell, D.; Senard, J.M.; Montastruc, J.L. A study of salivary secretion in Parkinson's disease. *Clin. Neuropharmacol.* **1999**, *22*, 213–215. [PubMed]
5. Proulx, M.; de Courval, F.P.; Wiseman, M.A.; Panisset, M. Salivary production in Parkinson's disease. *Mov. Disord.* **2005**, *20*, 204–207. [CrossRef] [PubMed]
6. Marg, S.; Walz, B.; Blenau, W. The effects of dopamine receptor agonists and antagonists on the secretory rate of cockroach (Periplaneta americana) salivary glands. *J. Insect Physiol.* **2004**, *50*, 821–830. [CrossRef] [PubMed]
7. Koga, T.; Kobashi, M.; Mizutani, M.; Tsukamoto, G.; Matsuo, R. Area postrema mediates gastric motor response induced by apomorphine in rats. *Brain Res.* **2003**, *960*, 122–131. [CrossRef]
8. Pazo, J.H.; Belforte, J.E. Basal ganglia and functions of the autonomic nervous system. *Cell Mol. Neurobiol.* **2002**, *22*, 645–654. [CrossRef]
9. Del Tredici, K.; Hawkes, C.H.; Ghebremedhin, E.; Braak, H. Lewy pathology in the submandibular gland of individuals with incidental Lewy body disease and sporadic Parkinson's disease. *Acta Neuropathol.* **2010**, *119*, 703–713. [CrossRef]
10. Ciucci, M.R.; Russell, J.A.; Schaser, A.J.; Doll, E.J.; Vinney, L.M.; Connor, N.P. Tongue force and timing deficits in a rat model of Parkinson disease. *Behav. Brain Res.* **2011**, *222*, 315–320. [CrossRef]
11. Russell, J.A.; Ciucci, M.R.; Hammer, M.J.; Connor, N.P. Videofluorographic assessment of deglutitive behaviors in a rat model of aging and Parkinson disease. *Dysphagia* **2013**, *28*, 95–104. [CrossRef]
12. Nóbrega, A.C.; Rodrigues, B.; Torres, A.C.; Scarpel, R.D.; Neves, C.A.; Melo, A. Is drooling secondary to a swallowing disorder in patients with Parkinson's disease? *Parkinsonism Relat. Disord.* **2008**, *14*, 243–245. [CrossRef]
13. Umemoto, G.; Tsuboi, Y.; Kitashima, A.; Furuya, H.; Kikuta, T. Impaired food transportation in Parkinson's disease related to lingual bradykinesia. *Dysphagia* **2011**, *26*, 250–255. [CrossRef]
14. Kalf, J.G.; Munneke, M.; van den Engel-Hoek, L.; de Swart, B.J.; Borm, G.F.; Bloem, B.R.; Zwarts, M.J. Pathophysiology of diurnal drooling in Parkinson's disease. *Mov. Disord.* **2011**, *26*, 1670–1676. [CrossRef]
15. Ali, G.N.; Wallace, K.L.; Schwartz, R.; DeCarle, D.J.; Zagami, A.S.; Cook, I.J. Mechanisms of oral-pharyngeal dysphagia in patients with Parkinson's disease. *Gastroenterology* **1996**, *110*, 383–392. [CrossRef]
16. Sung, H.Y.; Kim, J.S.; Lee, K.S.; Kim, Y.I.; Song, I.U.; Chung, S.W.; Yang, D.W.; Cho, Y.K.; Park, J.M.; Lee, I.S.; et al. The prevalence and patterns of pharyngoesophageal dysmotility in patients with early stage Parkinson's disease. *Mov. Disord.* **2010**, *25*, 2361–2368. [CrossRef]
17. Nienstedt, J.C.; Buhmann, C.; Bihler, M.; Niessen, A.; Plaetke, R.; Gerloff, C.; Pflug, C. Drooling is no early sign of dysphagia in Parkinson's disease. *Neurogastroenterol. Motil.* **2018**, *30*, e13259. [CrossRef]
18. Stanković, I.; Petrović, I.; Pekmezović, T.; Marković, V.; Stojković, T.; Dragašević-Mišković, N.; Svetel, M.; Kostić, V. Longitudinal assessment of autonomic dysfunction in early Parkinson's disease. *Parkinsonism Relat. Disord.* **2019**, *66*, 74–79. [CrossRef]
19. Sánchez-Martínez, C.M.; Choreño-Parra, J.A.; Placencia-Álvarez, N.; Nuñez-Orozco, L.; Guadarrama-Ortiz, P. Frequency and dynamics of non-motor symptoms presentation in Hispanic patients with Parkinson disease. *Front. Neurol.* **2019**, *10*, 1197. [CrossRef]
20. Durcan, R.; Wiblin, L.; Lawson, R.A.; Khoo, T.K.; Yarnall, A.J.; Duncan, G.W.; Brooks, D.J.; Pavese, N.; Burn, D.J.; The ICICLE-PD Study Group. Prevalence and duration of non-motor symptoms in prodromal Parkinson's disease. *Eur. J. Neurol.* **2019**, *26*, 979–985. [CrossRef]
21. Qin, X.; Li, X.; Xin, Z.; Li, Z. Gastrointestinal dysfunction in Chinese patients with Parkinson's disease. *Parkinson's Dis.* **2019**, *2019*, 3897315. [CrossRef]
22. Mao, C.J.; Xiong, Y.T.; Wang, F.; Yang, Y.P.; Yuan, W.; Zhu, C.; Chen, J.; Liu, C.F. Motor subtypes and other risk factors associated with drooling in Parkinson's disease patients. *Acta Neurol. Scand.* **2018**, *137*, 509–514. [CrossRef]
23. Ou, R.; Guo, X.; Wei, Q.; Cao, B.; Yang, J.; Song, W.; Chen, K.; Zhao, B.; Chen, X.; Shang, H. Diurnal drooling in Chinese patients with Parkinson's disease. *J. Neurol. Sci.* **2015**, *353*, 74–78. [CrossRef] [PubMed]
24. Fereshtehnejad, S.M.; Skogar, Ö.; Lökk, J. Evolution of Orofacial symptoms and disease progression in idiopathic Parkinson's disease: Longitudinal data from the Jönköping Parkinson registry. *Parkinson's Dis.* **2017**, *2017*, 7802819. [CrossRef] [PubMed]
25. Van Wamelen, D.J.; Leta, V.; Johnson, J.; Ocampo, C.L.; Podlewska, A.M.; Rukavina, K.; Rizos, A.; Martinez-Martin, P.; Chaudhuri, K.R. Drooling in Parkinson's disease: Prevalence and progression from the non-motor international longitudinal study. *Dysphagia* **2020**, *35*, 955–961. [CrossRef] [PubMed]
26. Karakoc, M.; Yon, M.I.; Cakmakli, G.Y.; Ulusoy, E.K.; Gulunay, A.; Oztekin, N.; Ak, F. Pathophysiology underlying drooling in Parkinson's disease: Oropharyngeal bradykinesia. *Neurol. Sci.* **2016**, *37*, 1987–1991. [CrossRef]
27. Rana, A.Q.; Yousuf, M.S.; Awan, N.; Fattah, A. Impact of progression of Parkinson's disease on drooling in various ethnic groups. *Eur. Neurol.* **2012**, *67*, 312–314. [CrossRef]
28. Reynolds, H.; Miller, N.; Walker, R. Drooling in Parkinson's disease: Evidence of a role for divided attention. *Dysphagia* **2018**, *33*, 809–817. [CrossRef]
29. Leclair-Visonneau, L.; Magy, L.; Volteau, C.; Clairembault, T.; Le Dily, S.; Préterre, C.; Peyre, A.; Damier, P.; Neunlist, M.; Péréon, Y.; et al. Heterogeneous pattern of autonomic dysfunction in Parkinson's disease. *J. Neurol.* **2018**, *265*, 933–941. [CrossRef]
30. Troche, M.S.; Okun, M.S.; Rosenbek, J.C.; Altmann, L.J.; Sapienza, C.M. Attentional resource allocation and swallowing safety in Parkinson's disease: A dual task study. *Parkinsonism Relat. Disord.* **2014**, *20*, 439–443. [CrossRef]

31. Zlotnik, Y.; Balash, Y.; Korczyn, A.D.; Giladi, N.; Gurevich, T. Disorders of the oral cavity in Parkinson's disease and parkinsonian syndromes. *Parkinson's Dis.* **2015**, *2015*, 379482. [CrossRef]
32. Ding, C.; Palmer, C.J.; Hohwy, J.; Youssef, G.J.; Paton, B.; Tsuchiya, N.; Stout, J.; Thyagarajan, D. Parkinson's disease alters multisensory perception: Insights from the rubber hand illusion. *Neuropsychologia* **2017**, *97*, 38–45. [CrossRef]
33. Bushmann, M.; Dobmeyer, S.M.; Leeker, L.; Perlmutter, J.S. Swallowing abnormalities and their response to treatment in Parkinson's disease. *Neurology* **1989**, *39*, 1309. [CrossRef]
34. Baijens, L.W.J.; Speyer, R.; Passos, V.L.; Pilz, W.; Roodenburg, N.; Clave, P. Swallowing in Parkinson patients versus healthy controls: Reliability of measurements in videofluoroscopy. *Gastroenterol. Res. Pract.* **2011**, *2011*, 380682. [CrossRef]
35. Leopold, N.A.; Kagel, M.C. Prepharyngeal dysphagia in Parkinson's disease. *Dysphagia* **1996**, *22*, 14–22. [CrossRef]
36. Meningaud, J.P.; Pitak-Arnnop, P.; Chikhani, L.; Bertrand, J.C. Drooling of saliva: A review of the etiology and management options. *Oral Surg. Oral Med. Oral Pathol. Oral Radiol. Endod.* **2006**, *101*, 48–57. [CrossRef]
37. Marinus, J.; van Hilten, J.J. The significance of motor (A)symmetry in Parkinson's disease. *Mov. Disord.* **2015**, *30*, 379–385. [CrossRef]
38. Pirker, W. Correlation of dopamine transporter imaging with parkinsonian motor handicap: How close is it? *Mov. Disord.* **2003**, *18*, S43–S51. [CrossRef]
39. Miller, N.; Walshe, M.; Walker, R.W. Sialorrhea in Parkinson's disease: Prevalence, impact and management strategies. *Res. Rev. Parkinsonism* **2019**, *9*, 17–28. [CrossRef]
40. Hou, Y.; Luo, C.; Yang, J.; Song, W.; Ou, R.; Liu, W.; Gong, Q.; Shang, H. A resting-state fMRI study on early-stage drug-naive Parkinson's disease patients with drooling. *Neurosci. Lett.* **2016**, *634*, 119–125. [CrossRef]
41. Schiffman, S.S.; Miletic, I.D. Effect of taste and smell on secretion rate of salivary IgA in elderly and young persons. *Physiol Behav.* **1996**, *60*, 243–248.
42. Alhajj, M.; Babos, M. Physiology, Salivation. [Updated 2020 Aug 11]. In *StatPearls [Internet]*; StatPearls Publishing: Treasure Island, FL, USA, 2021.
43. Nicaretta, D.H.; de Rosso, A.L.; Maliska, C.; Costa, M.M. Scintigraphic analysis of the parotid glands in patients with sialorrhea and Parkinson's disease. *Parkinsonism Relat. Disord.* **2008**, *14*, 338–341. [CrossRef]
44. Zhang, T.M.; Yu, S.Y.; Guo, P.; Du, Y.; Hu, Y.; Piao, Y.S.; Zuo, L.J.; Lian, T.H.; Wang, R.D.; Yu, Q.J.; et al. Nonmotor symptoms in patients with Parkinson disease: A crosssectional observational study. *Medicine* **2016**, *95*, e5400. [CrossRef] [PubMed]
45. Barbe, A.G.; Bock, N.; Derman, S.H.M.; Felsch, M.; Timmermann, L.; Noack, M.J. Self-assessment of oral health, dental health care and oral health-related quality of life among Parkinson's disease patients. *Gerodontology* **2017**, *34*, 135–143. [CrossRef] [PubMed]
46. Malek, N.; Lawton, M.A.; Grosset, K.A.; Bajaj, N.; Barker, R.A.; Burn, D.J.; Foltynie, T.; Hardy, J.; Morris, H.R.; Williams, N.M.; et al. Autonomic dysfunction in early Parkinson's disease: Results from the United Kingdom tracking Parkinson's study. *Mov. Disord. Clin. Pract.* **2017**, *4*, 509–516. [CrossRef] [PubMed]
47. Braak, H.; Del Tredici, K.; Rub, U.; de Vos, R.A.; Jansen Steur, E.N.; Braak, E. Staging of brain pathology related to sporadic Parkinson's disease. *Neurobiol. Aging* **2003**, *24*, 197–211. [CrossRef]
48. Cantuti-Castelvetri, I.; Keller-McGandy, C.; Bouzou, B.; Asteris, G.; Clark, T.W.; Frosch, M.P.; Standaert, D.G. Effects of gender on nigral gene expression and Parkinson disease. *Neurobiol. Dis.* **2007**, *26*, 606–614. [CrossRef] [PubMed]
49. Haaxma, C.A.; Bloem, B.R.; Borm, G.F.; Oyen, W.J.G.; Leenders, K.L.; Eshuis, S.; Booij, J.; Dluzen, D.E.; Horstink, M.W.I.M. Gender differences in Parkinson's disease. *J. Neurol. Neurosurg. Psychiatry* **2007**, *78*, 819–824. [CrossRef]
50. Conforti, R.; Capasso, R.; Negro, A.; Della Gatta, L.; De Cristofaro, M.; Amato, M.; Giganti, M.; Genovese, E.A. Diagnostic tools in neurodegenerative disorders of adult-elderly. *Recenti Prog. Med.* **2013**, *7*, 295–298.
51. Luchesi, K.F.; Kitamura, S.; Mourão, L.F. Dysphagia progression and swallowing management in Parkinson's disease: An observational study. *Braz. J. Otorhinolaryngol.* **2015**, *81*, 24–30. [CrossRef]
52. Morgan, J.; Sethi, K.D. Levodopa and the progression of Parkinson's disease. *Curr. Neurol. Neurosci. Rep.* **2005**, *5*, 261–262. [CrossRef]
53. David, N. Clinical features associated with drooling in Parkinson's disease. *Neurol. Sci.* **2021**, *42*, 895–903.
54. Merello, A.; Colosimo, C.; Hoffman, M.; Starkstein, S.; Leiguarda, R. Sleep benefit in Parkinson's disease. *Mov. Disord.* **1997**, *12*, 506–508. [CrossRef]
55. Rosenberg-Katz, K.; Herman, T.; Jacob, Y.; Giladi, N.; Hendler, T.; Hausdorff, J.M. Gray matter atrophy distinguishes between Parkinson disease motor subtypes. *Neurology* **2013**, *80*, 1476–1484. [CrossRef] [PubMed]
56. Selikhova, M.; Williams, D.R.; Kempster, P.A.; Holton, J.L.; Revesz, T.; Lees, A.J. A clinico-pathological study of subtypes in Parkinson's disease. *Brain* **2009**, *132*, 2947–2957. [CrossRef]
57. Cotzias, G.C.; Van Woert, M.H.; Schiffer, L.M. Aromatic amino acids and modification of Parkinsonism. *N. Engl. J. Med.* **1967**, *276*, 374–379. [CrossRef] [PubMed]
58. Calabresi, P.; Di Filippo, M.; Ghiglieri, V.; Tambasco, N.; Picconi, B. Levodopa-induced dyskinesias in patients with Parkinson's disease: Filling the bench-to-bedside gap. *Lancet Neurol.* **2010**, *9*, 1106–1117. [CrossRef]
59. Mito, Y.; Yabe, I.; Yaguchi, H.; Sato, C.; Takei, T.; Terae, S.; Tajima, Y. Relationships of drooling with motor symptoms and dopamine transporter imaging in drug-naïve Parkinson's disease. *Clin. Neurol. Neurosurg.* **2020**, *195*, 105951. [CrossRef]

60. Seibyl, J.P.; Marchek, K.L.; Quinlan, D.; Sheff, K.; Zoghbi, S.; Zea-Ponce, Y.; Baldwin, R.M.; Fussell, B.; Smith, E.O.; Charney, D.S.; et al. Decreased single-photon emission computed tomographic {123I}β-CIT striatal uptake correlates with symptom severity in parkinson's disease. *Ann. Neurol.* **1995**, *38*, 589–598. [CrossRef]
61. Benamer, H.T.; Patterson, J.; Wyper, D.J.; Hadley, D.M.; Macphee, G.J.; Grosset, D.G. Correlation of Parkinson's disease severity and duration with 123I-FP-CIT SPECT striatal uptake. *Mov. Disord.* **2000**, *15*, 692–698. [CrossRef]
62. Nascimento, D.; Carmona, J.; Mestre, T.; Ferreira, J.J.; Guimarães, I. Drooling rating scales in Parkinson's disease: A systematic review. *Parkinsonism Relat. Disord.* **2021**, *91*, 173–180. [CrossRef] [PubMed]
63. Proserpio, C.; de Graaf, C.; Laureati, M.; Pagliarini, E.; Boesveldt, S. Impact of ambient odors on food intake, saliva production and appetite ratings. *Physiol. Behav.* **2017**, *174*, 35–41. [CrossRef] [PubMed]
64. Hyson, H.C.; Johnson, A.M.; Jog, M.S. Sublingual atropine for sialorrhea secondary to Parkinsonism: A pilot study. *Mov. Disord.* **2002**, *17*, 1318–1320. [CrossRef]

healthcare

MDPI

Article

Effects of Cognitive Exercise Therapy on Upper Extremity Sensorimotor Function and Activities of Daily Living in Patients with Chronic Stroke: A Randomized Controlled Trial

Wonho Choi

Department of Physical Therapy, Gachon University, Incheon 21936, Korea; whchoi@gachon.ac.kr;
Tel.: +82-32-820-4423; Fax: +82-32-820-4420

Abstract: This study investigated the effects of cognitive exercise therapy on upper extremity sensorimotor function and daily activity in patients with chronic stroke and compared these effects to those of conventional occupational therapy. The 30 patients with chronic stroke (mean age: 63.6 ± 12.7 years; height: 162.8 ± 8.1 cm; weight: 60.6 ± 7.6 kg; body mass index: 22.8 ± 1.9 kg/m^2) were divided into two treatment groups with 15 patients in each. The respective interventions were provided for 30 min per day, five times weekly for 4 weeks. Manual and sensory function tests were conducted to evaluate the sensorimotor function, while the Korean-Modified Barthel Index was used to assess daily activities. All outcome variables were assessed before and after the interventions. A significant interaction was observed in sensory function ($p = 0.001$) but not motor function or daily activities ($p > 0.05$). No significant main group effects were found for any outcome variables ($p > 0.05$). The experimental group showed significant improvements in motor function ($p < 0.001$), sensory function ($p < 0.001$), and daily life activities ($p = 0.001$) after cognitive exercise therapy, whereas the control group showed significant improvement only in daily life activities post-intervention ($p = 0.012$). These results demonstrated the positive effects of cognitive exercise therapy on upper extremity sensorimotor function and daily life activities and the lack of improvement in motor and sensory function following conventional occupational therapy in patients with chronic stroke. Thus, the combination of cognitive exercise and conventional occupational therapies may be an effective way to improve sensory function and upper extremity motor function in patients with chronic stroke.

Keywords: cognitive exercise therapy; stroke; sensorimotor; activity of daily living

Citation: Choi, W. Effects of Cognitive Exercise Therapy on Upper Extremity Sensorimotor Function and Activities of Daily Living in Patients with Chronic Stroke: A Randomized Controlled Trial. *Healthcare* **2022**, *10*, 429. https://doi.org/10.3390/healthcare10030429

Academic Editors: Grigorios Nasios, Lambros Messinis, Efthimios Dardiotis and Markos Sgantzos

Received: 16 January 2022
Accepted: 23 February 2022
Published: 24 February 2022

1. Introduction

A stroke, a sudden impairment of body function caused by a blockage of blood flow to the brain, has the third-highest mortality rate after cancer and heart disease [1,2]. More than 85% of patients with stroke experience hemiplegia, and 55–75% of these patients have upper extremity impairment [3]. Impaired upper extremity function may result in decreased mobility of the shoulder joint, muscle weakness, sensory impairment, spasticity, and lack of coordination [4]. Subsequently, these lead to sequelae such as limited joint movement and limited upper extremity function [5]. Sensory impairment occurs in at least 50% of stroke patients and is expected to be higher with accurate examinations [6]. Sensory impairment interferes with correct movement and sharply reduces movement based on sensory input or feedback [7]. Patients with sensory impairment avoid movement, and their movements are dull and uncoordinated [7,8]. Thus, for patients with sensory impairment, proprioception, tactile sensation, pressure sensation, and stereoscopic sensation are essential for smooth movement by activating natural movements [9]. As these sensations are closely related to functional recovery after stroke, the impaired sensory function is one of the most significant factors hindering rehabilitation in patients with stroke. Additionally, impairment of upper extremity and sensory functions reduce the performance of basic daily activities. Therefore, patients cannot perform these activities independently and rely on the help of their caregivers [10].

Many interventions have been applied to address upper extremity function in patients with stroke. Among these, task-oriented upper extremity training improved the function of affected upper extremities [11–13]. The complex interaction between the left and right brains during bilateral tasks of the upper extremities enhanced the function of the affected upper extremity [11,12]. Moreover, constraint-induced movement therapy in the affected upper extremity of patients with stroke improved upper extremity function [13]. However, these interventions are therapeutically accessible only to patients with at least minimal voluntary motor skills. Another intervention is mirror therapy, which reorganizes brain areas along with physical rehabilitation. Based on the theoretical mirror neuron system, this intervention method promotes functional recovery of the upper extremities by inducing the recovery of motor function and movement on the affected side [14]. Another invention is mental practice, which allows patients to acquire and improve motor skills through thoughts of certain movements rather than actual body movements [15]. When a certain level of sensory recovery can be expected, repeated stimulation may provide necessary sensory feedback; moreover, the senses may be improved to a conscious level by focusing on a given sense [16].

Cognitive exercise therapy is a therapeutic intervention that aims to reorganize the central nervous system through learning motor function recovery [17,18]. This therapy emphasizes the close relationship of motor function to the activation of cognitive processes in the brain, such as perception, attention, memory, judgment, and language. The quality of recovery depends on the correct identification of the cognitive factors [19]. Cognitive exercise therapy allows various movements or actions to be performed through cognitive training processes, rather than movement training through interaction between the body and environment, to build a brain schema with four principles. The first principle is 'attention', which focuses on enhancing the effectiveness of treatment and reorganizing exercises in the treatment process. Second, patient treatment requires attention to somatosensory information with the eyes closed. Third, specific treatment instruments or tools are used to treat cognitive problems through interactions between the body and the environment. Finally, cognitive exercise does not force patients to conduct muscle contractions to accurately mobilize motor units. As such, the purpose of cognitive exercise therapy is not to teach patient-specific body postures but rather to develop and maximize the ability to organize the spatial, temporal, and intensity factors of the exercise sequence in the interaction between the body and environment [20].

Studies on cognitive exercise therapy in patients with stroke have shown improved upper extremity functions such as motor function, manual skills, and strength through image training of cognitive exercise therapy [21]. A comparison of cognitive exercise profiles showed improvements in sensory recognition and movement in the paralyzed arm [22]. In another study, activities of daily living (ADL) and motor function of the upper extremity improved after cognitive exercise therapy through contact tasks using a sponge on the paralyzed arm and spatial tasks using a graphic panel in patients with acute stroke. In addition, cognitive exercise profiling improved joint angle, spatiality, and shape recognition in recognition patterns [23].

However, most studies on cognitive exercise therapy were single-case or single-group studies with non-randomized groups and without a control group. Thus, these effects have not been directly compared. In addition, while various assessments have been conducted on upper extremity functions and daily activities, changes in cognitive patterns and pathological characteristics were described rather than used to directly assess sensory function recovery, which is a key factor in cognitive exercise therapy. While this approach is suitable for a single case study or qualitative research with a small number of subjects, it is difficult to quantify the objective results. Therefore, this study evaluated the effects of cognitive exercise therapy and compared them to conventional occupational therapy on upper extremity motor function, sensory function, and daily activities in patients with chronic stroke. We hypothesized that cognitive exercise therapy would improve upper extremity motor and sensory function and daily activities compared to conventional occupational therapy.

2. Materials and Methods

2.1. Ethical Approval

The study was conducted according to the guidelines of the Declaration of Helsinki and was approved by the Gachon University Institutional Review Board (IRB, 1044396-201804-HR-097-01). All participants signed a statement providing their informed consent before beginning the study. This trial was not registered. All procedures were conducted as approved by the IRB, and there was no selective outcome reporting.

2.2. Participants

A total of 37 patients with chronic stroke voluntarily participated in the study; among these, seven patients who did not meet the inclusion criteria were excluded. Thus, this study included 30 patients.

The inclusion criteria of the study participants were: patients diagnosed with hemiplegia due to stroke at 6 months after onset. The following types of patients were excluded: patients with cognitive impairment or dementia with a Korean version of mini-mental state examination (K-MMSE) score of 19 or higher who could understand and follow the instructions of the therapist; patients with Brunnstrom upper limb recovery stage 2 or higher; patients with a modified Ashworth scale stiffness level of 1+ or lower; patients with auditory and visual impairment or visual field defects; and patients with severe contracture due to orthopedic disease of the shoulder, elbow, and wrist joints.

2.3. Procedure

The 30 patients were randomly assigned to the control, and experimental groups administered conventional occupational and cognitive exercise therapies, respectively, with 15 patients assigned to each group using simple randomization methods that were independently conducted. Concealed allocation was performed using a computer-generated randomized table of numbers before data collection.

All participants in both groups underwent a manual function test (MFT), sensory function test (SFT), and Korean version of the modified Bathel index (K-MBI) evaluation for baseline assessment. The control group underwent conventional occupational therapy for 30 min twice daily, five times weekly, for 4 weeks (a total of 20 times) with the help of an experienced occupational therapist. The experimental group received conventional occupational therapy for 30 min per session, five times weekly, as well as cognitive exercise therapy for 30 min per session, five times a week, for 4 weeks (a total of 20 sessions). Both groups underwent their respective interventions for 30 min each in the morning and 30 min in the afternoon to minimize the physical fatigue (Figure 1).

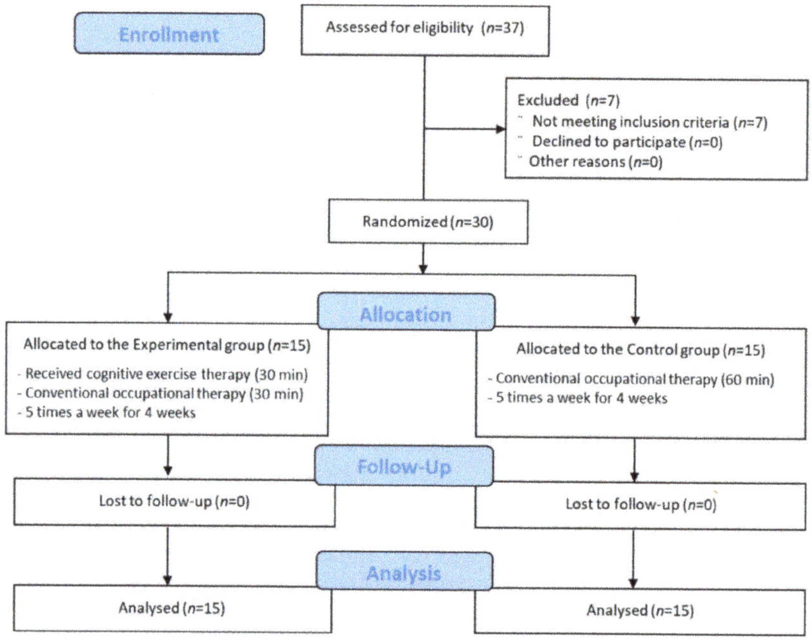

Figure 1. Flow diagram.

2.4. *Measurements*

2.4.1. Manual Function Test

The MFT was developed to measure the motor function of the affected upper limb in patients with a stroke [24]. The MFT comprises 32 test items that reflect the recovery of upper limb motor function and functional level of daily life activities. The reliability of the MFT was consistently above 0.95, and the validity of the MFT was good for both the Brunnstrom stage and the Stroke Impairment Assessment Set [25].

2.4.2. Sensory Function Test

The SFT examined seven sensations (pain, tactile, pressure, temperature, kinesthetic, stereognostic, and discriminative senses) in the hemiplegic upper limbs. The total score is 14 points for a total of seven sensations, with participants receiving 0 points they did not recognize both the sensory type and location of stimulation, 1 point for only one of the two, and 2 points for both sensation type and location. A preliminary test was performed three times on the unaffected upper limb before conducting the test for patients to understand the test. Ten tests were conducted on each item, and if more than seven of them were successful, the sensation was considered intact.

2.4.3. Korean-Modified Barthel Index

The K-MBI, which consists of 10 items describing activities of daily living (ADL) and mobility, was scored to measure the degree of assistance required by an individual and was used to assess ADL in patients with stroke [26]. Each item is rated 5-Likert scale, with weights added according to the item. The higher the total score, the more independent on performing ADLs [26]. The reported reliability of the K-MBI was 0.994, with a range of discriminative index of 0.783–0.909 [27].

2.5. Intervention

2.5.1. Cognitive Exercise Therapy

Cognitive exercise therapy involves two types of tasks: spatial and tactile. Spatial cognitive tasks involve perceiving elements as direction and distance, while tactile cognitive tasks involve perceiving properties from contact with an object, such as the surface, pressure, friction, and weight. Each participant in this study performed five spatial and tactile tasks. Two or three tasks were selected based on the participants' recovery levels.

Shoulder joint recognition training by motor imagery [28]: the spatial task in shoulder joint recognition training is to control the shoulder joint through sensory images of the joint during spatial tasks. The images help improve recognition movements during bending and opening the shoulder joint. The participants were asked to sit upright on a chair without touching the backrest. After bending and opening the shoulder joint on the unaffected side, the participants were asked to describe the sensation. The therapist passively moved the participant's shoulder joint on the affected side and asked the participants to describe the sensation again. Then, motor imagery of the unaffected side was used to induce movement of the unaffected side. The participants were asked to describe the sensation of the shoulder joint on the affected side. The feeling of bending and opening of the shoulder joint on the affected side was expressed with active assistance from the therapist. The feelings of the shoulder joint's movement on the affected and unaffected sides were compared. The participants were then asked to express their sensory images during the movement of the shoulder joint on the unaffected and affected sides. The images were compared, and the differences were expressed in words. This training increased participant awareness of the shoulder joint using the movement image of the shoulder joint (Figure 2A).

Shoulder and elbow joint recognition training using a circular track plate [19]: shoulder and elbow joint recognition training, a cognitive task that uses special senses, was conducted using a circular track plate to distinguish the movement angle of the shoulder joint. In this task, the participants distinguished the distance of each circle of the track place to improve recognition of shoulder joint movement. The participants were asked to sit on a chair without touching the backrest in front of the track plate on a table. The size of each track plate was analyzed visually. After blinding the participant, the therapist supported the participant's arm to provide passive movement, which provided information on the circle distances and locations. Each participant was given a cognitive task to identify the distance and describe the response in words. After unblinding, the participants' descriptions were compared to the results of the visual analysis. The participants perceived cognitive tasks to identify distance through movements of the shoulder joint in space (Figure 2B).

Elbow and wrist joint angle awareness training using a Bogen [19]: among spatial tasks, this training on distance identified the movement angle of the elbow joint using a Bogen. This task was conducted to distinguish the distance to each grid line of Bogen and increase awareness of the elbow joint or wrist movement. The Bogen was placed on a table, and the participant was asked to sit in front of the table on a chair without touching the backrest. The participants visually observed the differences in grid lines and were blinded afterward. The elbow joint was then placed on the table, and the therapist supported the patient's forearm and fingers to help the participant passively identify differences in distance. Afterward, each participant was given a cognitive task to identify distance and was asked to describe their response in words. After unblinding, the participants' descriptions were compared with the visual analysis results. The task was gradually subdivided using a Bogen with finer grids to adjust the difficulty (Figure 2C).

Elbow and wrist pressure awareness training using a sponge [22]: this tactile task involved the use of the shoulder, elbow, and wrist joints to perceive the pressure differences. Sponges were used to improve the ability to perceive the gradient pressure differences. The participants were asked to sit on a chair without touching the backrest, and three types of sponges of different thicknesses were used. After blinding the participants, the therapist placed a sponge on one area of the body and gently pressed the sponge. The participants were asked to pay attention to the area in which the sponge was placed and to distinguish differences in the hardness of the sponge. During this process, the participant was asked to pay attention to the body part (elbow, shoulder, or wrist) rather than the sponge. The therapist asked for differences in the position and hardness of the sponges. If the participant was able to distinguish the differences in the hardness of the sponge in one area on the unaffected side, the pressure differences between the affected and unaffected sides were assessed. The task was conducted with the participant in a relaxed state, and the task allowed the participant to increase recognition of the body by distinguishing differences in sponge pressure (Figure 2D).

Finger tactile recognition training using a tactile plate [19]: this tactile task involved distinguishing between materials using the fingers and a tactile plate for the recognition of surface materials and perception of finger movements. Each participant was asked to sit on a chair without touching the backrest and to place both arms on a table. Five tactile plates of different soft or rough materials were prepared. Visual and tactile information for both the affected and unaffected sides was provided to the participant. The therapist passively moved the participant's finger to provide tactile sensations to the plates, and the differences in the surface materials and finger movements were evaluated. The tactile sensation was further tested using plates made of materials with minor differences (Figure 2E).

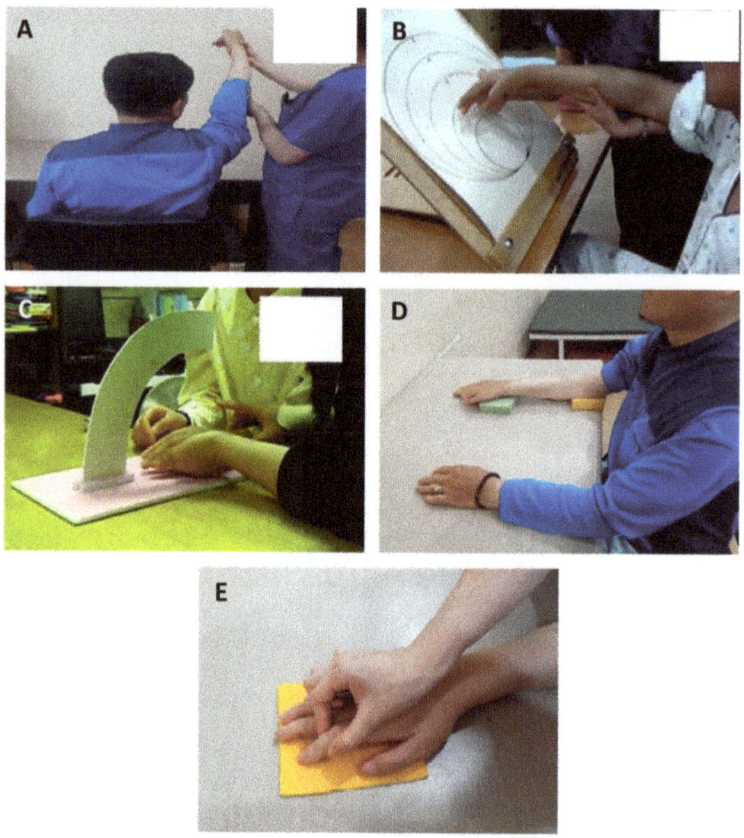

Figure 2. (**A**) Shoulder joint recognition training by motor imagery. (**B**) Shoulder and elbow joint recognition training using a circular track plate. (**C**) Training on awareness of elbow and wrist joint angles using a Bogen. (**D**) Training on pressure awareness of the elbow and wrist using a sponge. (**E**) Finger tactile recognition training using a tactile plate.

2.5.2. Conventional Occupational Therapy

The conventional occupational therapy provided in this study was based on the occupational therapy practice framework (OTPF3) [29]. The therapy included passive joint exercises to reduce the spasticity of the upper extremity muscles and joints affected by stroke, as well as active joint exercises such as small muscle activity using different tools, micro-movement training, and two-handed exercises. Neurodevelopmental treatment and pain management were provided according to the patient's level of recovery. ADL, including wheelchair operation, climbing up and down stairs, functional movement, eating, dressing, and personal hygiene, were assessed. Depending on the level of recovery of the participant, 2–3 tasks were selected and performed.

2.6. Sample Size Estimation

The sample size was estimated using G*power 3.1.9.4 (Heinrich Heine University, Dusseldorf, Dusseldorf, Germany). An estimated sample size of 26 participants was obtained using an effect size of 1.03, the alpha error probability of 0.05, and power of 0.80 for the effect of cognitive exercise therapy on upper limb function when a clinically significant difference was observed between two independent means with a one-tailed test. The effect size (Cohen's d) of 1.03 was calculated from the results of a previous study [20]. An additional 10% of participants were recruited to account for unanticipated attrition.

2.7. Statistical Analysis

Statistical analysis was performed using IBM SPSS Statistics for Windows, version 26.0 (IBM Corp., Armonk, NY, USA), and the measured values of all items were calculated as mean and standard deviation (SD). Shapiro–Wilk tests were used to test the normal distribution of the data. Chi-squared and independent t-tests were performed to compare the general participant characteristics using the homogeneity test between groups. Repeated-measures analysis of variance (RM ANOVA) was conducted to assess the group-by-time interaction effect of interest. When a significant interaction was found, a paired t-test was conducted to compare the outcome variables before and after the intervention in each group. The level of statistical significance was set at $\alpha = 0.05$.

3. Results

The general participant characteristics are shown in Table 1. These characteristics did not differ significantly between the groups ($p > 0.05$).

Table 1. General participant characteristics ($n = 30$).

	EG ($n = 15$)	CG ($n = 15$)	p
Age (years)	62.5 ± 11.3	64.7 ± 14.3	0.633
Sex, Female, n (%)	6 (40.0)	11 (73.3)	0.139
Height (cm)	165.2 ± 7.7	160.5 ± 8.0	0.118
Weight (kg)	61.5 ± 7.1	59.6 ± 8.2	0.493
BMI (kg/m^2)	22.5 ± 1.7	23.1 ± 2.2	0.456
Stroke type			
Infarction, n (%)	7 (46.7)	10 (66.7)	0.285
Hemorrhage, n (%)	8 (53.3)	5 (33.3)	
Affected side, left, n (%)	8 (53.3)	9 (60.0)	1.000
Onset (months)	19.9 ± 7.1	18.3 ± 7.0	0.540
K-MMSE (scores)	24.1 ± 3.8	23.9 ± 3.2	0.918

Abbreviations: EG—experimental group (cognitive exercise therapy); CG—control group (conventional occupational therapy); BMI—body mass index; K-MMSE—Korean version of Mini-Mental State Examination.

Comparisons of homogeneity between the control and experimental groups before their respective interventions showed no significant differences in MFT, SFT, and K-MBI at baseline ($p > 0.05$).

Table 2 shows the outcome variables before and after the interventions in the cognitive exercise therapy and conventional occupational therapy groups. There was no significant interaction in the mean MFT score between groups and time ($p = 0.626$). The main effect of the group was not significant ($p = 0.961$). The baseline MFT score, 8.40 ± 9.34 points, significantly improved to 9.93 ± 9.53 points post-intervention in the experimental group ($p < 0.001$), while the control group showed no significant improvement (8.80 ± 8.93 to 9.87 ± 9.61, $p > 0.05$).

Table 2. Motor and sensory functions and daily activity before and after the interventions in the groups ($n = 30$).

Outcome Variables		EG ($n = 15$)	CG ($n = 15$)	Group × Time Interaction F(p)	Main Effect of Group F(p)
MFT	Pre-test	8.40 ± 9.34	8.80 ± 8.93	0.242 (0.626)	0.002 (0.961)
	Post-test	9.93 ± 9.53	9.87 ± 9.61		
	t(p) *	<0.001	0.251		
SFT	Pre-test	7.07 ± 2.91	8.20 ± 3.32	13.225 (0.001)	0.114 (0.738)
	Post-test	8.87 ± 2.97	8.47 ± 2.88		
	t(p) *	<0.001	0.104		
K-MBI	Pre-test	57.20 ± 19.27	46.73 ± 22.59	0.728 (0.401)	1.627 (0.213)
	Post-test	61.47 ± 20.64	53.07 ± 19.42		
	t(p) *	0.001	0.012		

Abbreviations: EG—experimental group (cognitive exercise therapy); CG—control group (conventional occupational therapy); MFT—manual function test; SFT—sensory function test; K-MBI—Korean version of modified Barthel index. * within-group effect.

While a significant group-by-time interaction effect was observed for the mean SFT score ($p = 0.001$), no significant main effect of the group was observed ($p = 0.738$). The SFT improved significantly from 7.07 ± 2.91 points at baseline to 8.87 ± 2.97 points post-intervention in the experimental group ($p < 0.001$). Conventional occupational therapy also showed a significant improvement in SFT (8.20 ± 3.32 8.47 ± 2.88, $p = 0.104$). Subdivision of the SFT score showed no significant interactions for all sensations ($p > 0.05$), except for tactile and kinesthetic senses ($p < 0.001$ and 0.001, respectively).

No significant interaction was observed in the K-MBI score between group and time ($p = 0.401$), and no statistically significant difference was found ($p > 0.05$). Both groups showed significant improvements in daily activity before and after the intervention (57.20 ± 19.27 to 61.47 ± 20.64, $p = 0.001$ and 46.73 ± 22.59 to 53.07 ± 19.42, $p = 0.012$, respectively).

4. Discussion

A stroke interrupts the blood supply to the brain and causes bleeding in the brain tissue, leading to various types of neurological damage [30]. Motor weakness of the upper extremities is commonly observed after the stroke, which leads to independence in activities of daily living [31]. This damage affects various sensory elements that detect changes in movement or distinguish the direction of movement of the extremities. Therefore, functional exercises must be developed to increase the use of the upper extremities [20]. Cognitive exercise therapy controls the recovery of motor function in the somatosensory system by reproducing the interaction between the body and the environment in the cranial nerves and organizing activities in patients with disabilities. This therapy allows for the extensive recovery of patients from injuries [32]. However, most studies on cognitive exercise therapy have been single-case or group studies, including non-randomized groups without a control group. Therefore, the present study evaluated the effects of cognitive exercise therapy with spatial and tactile tasks for the shoulder, elbow, and wrist joints on upper extremity function, sensory function, and daily activity compared to those of conventional occupational therapy in patients with stroke.

The results of the current study showed significantly improved daily activities, upper extremity motor function, and sensory function among patients in the cognitive exercise therapy group, while patients in the conventional occupational therapy group did not show significantly improved motor and sensory function, despite the significant group differences. Consistent with our findings, interventions involving tactile and spatial task training in cognitive exercise therapy improved upper extremity and sensory function [21,23,33]. Ahn (2009) performed training to perceive the hardness of a sponge on the shoulder joint of the affected upper extremity, as well as spatial tasks to improve direction in the shoulder, elbow, and wrist joints. The author reported significantly improved upper extremity function of the affected side compared to that of the unaffected side [23]. In addition, the cognitive exercise profile showed significantly improved direction of motion and awareness of the joint angles on the affected side. This is consistent with our finding of improved upper extremity motor and sensory functions, as well as significantly improved sensorimotor function, in the cognitive exercise therapy group. Furthermore, a previous study that investigated the effects of tactile and spatial cognitive training on the reduction in stiffness in hemiplegic patients reported that pressure awareness training of the elbow joint using sponges and spatial awareness training using a circular track plate reduced arm muscle stiffness by 23.3% and improved upper extremity function by 9.37% [33]. In addition, cognitive-motor profiling showed improved angle and spatial awareness in the shoulder, elbow, and wrist joints. Finally, in their study on hemineglect patients, Lim and Lee (2014) showed that cognitive exercise therapy led to positive improvements in upper extremity motor function and grip test results, suggesting the possible positive effects of cognitive exercise therapy on the rehabilitation of patients with stroke and hemineglect [34].

Although we observed no significant group difference in SFT, there was a significant group effect in tactile sensation among the sub-areas in the SFT. Previous research showed that cognitive exercise therapy improved tasks performed using spatial and tactile senses and that activating the premotor cortex areas, which supports the improvement in tactile sensory functions of the hand observed in the present study [35]. While upper extremity function improved due to the recovery of tactile sensation, the MFT did not differ significantly following conventional occupational therapy, as the upper extremities are the last to improve after stroke [36].

We observed significantly improved sensorimotor function after the intervention in the experimental group but not in the control group. However, daily life activity significantly improved after the intervention in both groups, although the two groups did not differ significantly. A study that applied spatial and tactile tasks to patients with stroke reported findings contrary to those of previous studies, in which daily life activity was significantly improved compared to the control group [37]. Consistent with our results, other studies have also reported functional improvement in the upper extremities and daily life activity performance after cognitive exercise therapy.

Therefore, to complement the limitations of previous studies, this study enrolled more participants and conducted a total of 20 cognitive exercise therapy sessions for 4 weeks, including an SFT, which is a key part of cognitive exercise therapy and is important for functional recovery in patients with stroke. Both cognitive exercise and conventional occupational therapies had positive effects on ADLs in patients with stroke. In contrast, sensorimotor function recovery was significantly improved only in the cognitive exercise therapy group, although significant differences were observed between the groups. Lee (2016) examined changes in electroencephalography (EEG), which records brain activities when healthy individuals distinguish between regular and irregular shapes, to show that tactile stimulation activated the parietal region of the brain following cognitive exercise therapy compared to conventional occupational therapy [38]. Instead, attention and perceptual awareness indicate changes in the central nervous system. In other words, when participants paid close attention to the body to interpret somatosensory information, attention, recognition, and cognition to solve spatial and tactile tasks led to EEG changes in the cerebral cortex, which mediates upper extremity functions. Cognitive exercise therapy is thought to improve body part recognition in patients with impaired upper extremity function due to stroke [39]. Considering the body as a surface receptor and transmission of delicate information to the brain by the body may have led to the superficial senses of touch and pressure as well as deep senses of joints and muscles affecting sensorimotor function and ADLs [39].

The results of this study showed that cognitive exercise therapy had positive effects on upper extremity function and sensorimotor function in patients with chronic stroke, while conventional occupational therapy did not, suggesting that the brain is highly activated when perceiving information or paying attention during the spatial and tactile tasks in the cognitive intervention [40]. Although the present study objectively examined the function of eight sensors using the SFT, several limitations must be considered when interpreting the findings. First, there were limitations in providing an optimal environment to achieve the full focus of the participants during the interventions. The participants were also not followed-up for post-intervention effects; thus, the long-term effects of cognitive exercise therapy could not be assessed. Additional studies are needed to confirm that cognitive exercise therapy is an effective treatment with potential for further development and implications. In addition, although validated measurement tools were used, the reliability of the test was not measured in the present study, which may mask the true effects of the intervention. Lastly, since the cognitive exercise group received both cognitive exercise and conventional occupational therapies, the intervention benefit cannot be assumed to be only due to cognitive exercise therapy; therefore, further studies should include a true experimental group receiving only cognitive exercise.

5. Conclusions

In conclusion, both groups showed improvement in daily activities after the intervention. Additionally, cognitive exercise therapy using spatial and tactile tasks significantly improved upper extremity motor and sensory functions compared to conventional occupational therapy. Therefore, cognitive exercise therapy compensates for the shortcomings of conventional occupational therapy used for rehabilitation and provides an effective rehabilitation protocol for patients with chronic stroke.

Funding: This study received no external funding.

Institutional Review Board Statement: The study was conducted in accordance with the guidelines of the Declaration of Helsinki and approved by the Institutional Review Board of Gachon University (1044396-201804-HR-097-01).

Informed Consent Statement: Informed consent was obtained from all the participants involved in the study.

Data Availability Statement: The datasets generated during the current study are available from the corresponding author upon reasonable request.

Acknowledgments: The author would like to express gratitude to Joo-Young Park for helping with the data collection and analysis.

Conflicts of Interest: The authors declare no conflict of interest.

References

1. Lopez, A.D.; Mathers, C.D.; Ezzati, M.; Jamison, D.T.; Murray, C.J. Global and regional burden of disease and risk factors, 2001: Systematic analysis of population health data. *Lancet* **2006**, *367*, 1747–1757. [CrossRef]
2. Towfighi, A.; Saver, J.L.J.S. Stroke declines from third to fourth leading cause of death in the United States: Historical perspective and challenges ahead. *Stroke* **2011**, *42*, 2351–2355. [CrossRef]
3. Wolf, S.L.; Catlin, P.A.; Ellis, M.; Archer, A.L.; Morgan, B.; Piacentino, A.J.S. Assessing Wolf motor function test as outcome measure for research in patients after stroke. *Stroke* **2001**, *32*, 1635–1639. [CrossRef]
4. Pellegrino, L.; Coscia, M.; Muller, M.; Solaro, C.; Casadio, M. Evaluating upper limb impairments in multiple sclerosis by exposure to different mechanical environments. *Sci. Rep.* **2018**, *8*, 2110. [CrossRef]

5. Cirstea, M.; Ptito, A.; Levin, M.F. Arm reaching improvements with short-term practice depend on the severity of the motor deficit in stroke. *Exp. Brain Res.* **2003**, *152*, 476–488. [CrossRef]
6. Doyle, S.; Bennett, S.; Fasoli, S.E.; McKenna, K.T. Interventions for sensory impairment in the upper limb after stroke. *Cochrane Database Syst. Rev.* **2010**, *6*, CD006331. [CrossRef]
7. Patel, N.; Jankovic, J.; Hallett, M.J.T.L.N. Sensory aspects of movement disorders. *Lancet Neurol.* **2014**, *13*, 100–112. [CrossRef]
8. Kamalakannan, S.; Chockalingam, M.; Sethuraman, L.; Moorthy, S.D.; Muthuvel, T. Occupational Therapy for Reducing Disabilities in Persons with Disabilities in India: A Systematic Review. *J. Occup. Ther.* **2021**, *53*, 4. [CrossRef] [PubMed]
9. Schwartz, J.H.; Jessell, T.M. *Principles of Neural Science*; McGraw-Hill: New York, NY, USA, 2000.
10. Yoo-Soon, B. The Effects of Task-Oriented Activities on the Cognitive Function and Performance of Activities of Daily Living in Stroke Patients. *Korean J. Occup. Ther.* **2007**, *15*, 49–61.
11. Sabaté, M.; González, B.; Rodríguez, M.J.N. Brain lateralization of motor imagery: Motor planning asymmetry as a cause of movement lateralization. *Neuropsychologia* **2004**, *42*, 1041–1049. [CrossRef] [PubMed]
12. Jung, J.H.; Cho, Y.N.; Chae, S.Y. The Effect of Task-Oriented Movement Therapy on Upper Extremity, Upper Extremity Function and Activities of Daily Living for Stroke Patients. *J. Rehabil. Res.* **2011**, *15*, 231–253.
13. Jung, M.Y.; Chung, B.I. The Effect of Constraint-induced Movement Therapy on the Affected Upper Extremity Function and Activities of Daily Living for Stroke Patients. *Korean J. Occup. Ther.* **2009**, *17*, 77–90.
14. Sütbeyaz, S.; Yavuzer, G.; Sezer, N.; Koseoglu, B.F. Mirror therapy enhances lower-extremity motor recovery and motor functioning after stroke: A randomized controlled trial. *Arch. Phys. Med. Rehabil.* **2007**, *88*, 555–559. [CrossRef] [PubMed]
15. Richardson, A. Physical Education; Recreation. Mental practice: A review and discussion part I. *Res. Q. Am. Assoc. Health Phys. Educ. Recreat.* **1967**, *38*, 95–107.
16. Aman, J.E.; Elangovan, N.; Yeh, I.L.; Konczak, J. The effectiveness of proprioceptive training for improving motor function: A systematic review. *Front. Hum. Neurosci.* **2015**, *8*, 1075. [CrossRef]
17. Takeuchi, N.; Izumi, S.-I. Rehabilitation with poststroke motor recovery: A review with a focus on neural plasticity. *Stroke Res. Treat.* **2013**, *2013*, 128641. [CrossRef]
18. Franca, P. Introduction to neurocognitive rehabilitation according to Carlo Perfetti. *Vopr. Reabil.* **2018**, *3*, 10–21.
19. Jang, J.-Y.; Kim, J.-M. The Effects of Cognitive Therapy Exercise on the Affected Upper Extremity Function and Activities of Daily Living for Stroke Patient. *J. Korean Soc. Neurocogn. Rehabil.* **2012**, *4*, 55–65.
20. Lee, S.; Bae, S.; Jeon, D.; Kim, K.Y. The effects of cognitive exercise therapy on chronic stroke patients' upper limb functions, activities of daily living and quality of life. *J. Phys. Ther. Sci.* **2015**, *27*, 2787–2791. [CrossRef]
21. Jeong, E.-M.; Lee, E.-M.; Kim, S.-G. The Effect of Motor Image on the Upper Extremity Function Recovery in Stroke Patients. *J. Korean Soc. Neurocogn. Rehabil.* **2016**, *8*, 25–31.
22. Kim, Y.-S. Effect of Cognitive Therapeutic Exercise on Improvement Hand Function with Hemiplegic Patient After Stroke. *J. Korean Soc. Neurocogn. Rehabil.* **2009**, *1*, 11–24.
23. Ahn, S.-N.; Lee, J.-W. Effect of Cognitive Therapeutic Exercise on Recovery of the Upper Limb Function in hemiplegia. *J. Korean Soc. Neurocogn. Rehabil.* **2009**, *1*, 43–55.
24. Nakamura, R.; Moriyama, S.; Yamada, Y.; Seki, K. Recovery of impaired motor function of the upper extremity after stroke. *Tohoku J. Exp. Med.* **1992**, *168*, 11–20. [CrossRef] [PubMed]
25. Miyamoto, S.; Kondo, T.; Suzukamo, Y.; Michimata, A.; Izumi, S. Reliability and validity of the Manual Function Test in patients with stroke. *Am. J. Phys. Med. Rehabil.* **2009**, *88*, 247–255. [CrossRef] [PubMed]
26. Jung, H.Y.; Park, B.K.; Shin, H.S.; Kang, Y.K.; Pyun, S.B.; Paik, N.J.; Kim, S.H.; Kim, T.H.; Han, T.R. Development of the Korean version of Modified Barthel Index (K-MBI): Multi-center study for subjects with stroke. *J. Korean Acad. Rehabil. Med.* **2007**, *31*, 283–297.
27. Choi, Y.-I.; Kim, W.-H.; Park, E.-Y.; Kim, E.-J. The validity, reliability and discriminative index of the Korean version of Modified Barthel Index (K-MBI) in stroke patients. *J. Korea Acad. Coop. Soc.* **2012**, *13*, 4119–4125.
28. Choi, J.-E.; Lee, S.-S. Effect of Stretch Reflex Control of the Upper Limb on Recovery Hand Function with Stroke Patient. *J. Korean Soc. Neurocogn. Rehabil.* **2009**, *1*, 75–86.
29. Bier, F.O.; Fenn, E.O.; Kochobay, A.O.; Moran, K.O.; Naik, K.O. Occupational Therapy Practice Framework: Domain and Process (3rd Edition). *Am. J. Occup. Ther.* **2014**, *68*, S1–S48. [CrossRef]
30. Kuriakose, D.; Xiao, Z. Pathophysiology and Treatment of Stroke: Present Status and Future Perspectives. *Int. J. Mol. Sci.* **2020**, *21*, 7609. [CrossRef]
31. Raghavan, P. Upper Limb Motor Impairment After Stroke. *Phys. Med. Rehabil. Clin. N. Am.* **2015**, *26*, 599–610. [CrossRef]
32. Perfetti, C.; Wopfner-Oberleit, S. *Der hemiplegische Patient: Kognitiv therapeutische Übungen*; Pflaum: München, Germany, 1997.
33. Kim, M.-Y.; Lee, T.-Y. Effect of Tactile and Spacious Cognitive Task Training on Decrease of Spasticity With Hemiplegic Patient. *J. Korean Soc. Neurocogn. Rehabil.* **2010**, *2*, 56–66.
34. Lim, Y.-J.; Lee, S.-A. The Effect of Neurocognitve Rehabilitation on the Visual Perception and Upper-limb Function of Neglect. *J. Korean Soc. Neurocogn. Rehabil.* **2014**, *6*, 21–29.
35. Van de Winckel, A.; Sunaert, S.; Wenderoth, N.; Peeters, R.; Van Hecke, P.; Feys, H.; Horemans, E.; Marchal, G.; Swinnen, S.P.; Perfetti, C.J.N. Passive somatosensory discrimination tasks in healthy volunteers: Differential networks involved in familiar versus unfamiliar shape and length discrimination. *NeuroImage* **2005**, *26*, 441–453. [CrossRef] [PubMed]

36. Fritz, J.; Shamma, S.; Elhilali, M.; Klein, D. Rapid task-related plasticity of spectrotemporal receptive fields in primary auditory cortex. *Nat. Neurosci.* **2003**, *6*, 1216–1223. [CrossRef]
37. Kim, Y. The Effectiveness of Cognitive Therapeutic Exercise on Upper Extremity Function and Activities of Daily living in Stroke Patients. *J. Korean Soc. Neurocogn. Rehabil.* **2014**, *6*, 41–46.
38. Lee, J.; Jung, B.R.; Ahn, S.N. The Difference of Electronencephalograph on Recognition of Formal and Geometric Shape in Adults. *J. Korean Soc. Neurocogn. Rehabil.* **2016**, *8*, 1–7.
39. Hatem, S.M.; Saussez, G.; Della Faille, M.; Prist, V.; Zhang, X.; Dispa, D.; Bleyenheuft, Y. Rehabilitation of Motor Function after Stroke: A Multiple Systematic Review Focused on Techniques to Stimulate Upper Extremity Recovery. *Front. Hum. Neurosci.* **2016**, *10*, 442. [CrossRef]
40. Andersson, R.; Mohme-Lundholm, E. Studies on the relaxing actions mediated by stimulation of adrenergic alpha- and beta-receptors in taenia coli of the rabbit and guinea pig. *Acta Physiol. Scand.* **1969**, *77*, 372–384. [CrossRef]

healthcare

MDPI

Article

Cost-Effectiveness of Upper Extremity Dry Needling in Chronic Stroke

Daniel Fernández-Sanchis [1], Natalia Brandín-de la Cruz [1], Carolina Jiménez-Sánchez [1], Marina Gil-Calvo [2,3], Pablo Herrero [2,3,*] and Sandra Calvo [2,3]

1 Faculty of Health Sciences, Universidad San Jorge, Autovía A-23 Zaragoza-Huesca Km.299, Villanueva de Gállego, 50830 Zaragoza, Spain; efernandez@usj.es (D.F.-S.); nbrandin@usj.es (N.B.-d.l.C.); cjimenez@usj.es (C.J.-S.)

2 IIS Aragon, University of Zaragoza, Avda. San Juan Bosco, 13, 50009 Zaragoza, Spain; magilcal@unizar.es (M.G.-C.); sandracalvo@unizar.es (S.C.)

3 Department of Physiatry and Nursing, Faculty of Health Sciences, University of Zaragoza, C/Domingo Miral s/n., 50009 Zaragoza, Spain

* Correspondence: pherrero@unizar.es

Abstract: Introduction: Dry needling is a non-pharmacological approach that has proven to be effective in different neurological conditions. Objective: The aim of this study was to evaluate the cost-effectiveness of a single dry needling session in patients with chronic stroke. Methods: A cost-effectiveness analysis was performed based on a randomized controlled clinical trial. The results obtained from the values of the EuroQol-5D questionnaire and the Modified Modified Ashworth Scale were processed in order to obtain the percentage of treatment responders and the quality-adjusted life years (QALYs) for each alternative. The cost analysis was that of the hospital, clinic, or health center, including the equipment and physiotherapist. The cost per respondent and the incremental cost-effectiveness ratio of each alternative were assessed. Results: Twenty-three patients with stroke were selected. The cost of DN treatment was EUR 14.96, and the data analysis showed a favorable cost-effectiveness ratio of both EUR/QALY and EUR/responder for IG, although the sensitivity analysis using limit values did not confirm the dominance (higher effectiveness with less cost) of the dry needling over the sham dry needling. Conclusions: Dry needling is an affordable alternative with good results in the cost-effectiveness analysis—both immediately, and after two weeks of treatment—compared to sham dry needling in persons with chronic stroke.

Keywords: cost–utility; stroke; upper extremity; EQ-5D

Citation: Fernández-Sanchis, D.; Brandín-de la Cruz, N.; Jiménez-Sánchez, C.; Gil-Calvo, M.; Herrero, P.; Calvo, S. Cost-Effectiveness of Upper Extremity Dry Needling in Chronic Stroke. *Healthcare* **2022**, *10*, 160. https://doi.org/10.3390/healthcare10010160

Academic Editors: Grigorios Nasios, Lambros Messinis, Efthimios Dardiotis and Markos Sgantzos

Received: 13 December 2021
Accepted: 12 January 2022
Published: 14 January 2022

1. Introduction

Stroke is a major contributor to disability worldwide, and the second leading cause of death in Spain [1,2], generating a great impact on patients' quality of life (QOL) due to the functional limitations that it entails [3]. According to the Global Burden of Disease Study (GBD), the socioeconomic burden of stroke has increased over time, although there has been a decrease in its prevalence [2]. Stroke imposes a high burden in terms of direct and indirect costs: on the one hand, indirect costs because of lost productivity due to patients' long-term disability, restricted social functioning, and premature death, leading to a detriment to the patients' quality of life; on the other hand, direct costs of care resulting from costs of health professionals, hospital services, medications, etc. [4].

Upper motor neuron lesions may result in long-term positive and/or negative symptoms [5], which usually lead to different degrees of upper extremity disability. Spasticity is one of the more common symptoms, leading to progressive functional limitation and decrease in quality of life [6]. The current scientific evidence shows the effectiveness of different interventions for the rehabilitation of upper extremity post-stroke, such as robotics [7], virtual reality [8], and different physical therapy protocols for functional improvement [9]. Physical therapy treatments can be combined with other pharmacological interventions and/or other medical treatments, such as antispastic drugs or botulinum toxin type A (BTX-A) infiltration [10]. Recently, non-pharmacological approaches have been used, such as dry needling (DN) of myofascial trigger points, which is increasingly used to treat neurological conditions such as stroke [11–14], Parkinson´s disease [15], and multiple sclerosis [16]. Although the reasons for the increase in non-pharmacological treatments such as DN are not clear, the following factors could be relevant: (1) from the patient´s perspective, the adverse effects of pharmacological treatments, or a shift to more patient-centered treatments, where patients are more involved in decision making on different treatment alternatives; and (2) from the professional and health system perspective, the high costs of pharmacological treatments such as BTX-A infiltration.

In relation to DN and BTX-A, BTX-A is the most potent neurotoxin known, and its paralytic effect is due to the blockade of neuromuscular transmission [17]. On the other hand, DN acts by mechanically impairing the sensory or motor components of nerve endings and dysfunctional motor endplates that contribute to the abnormal functioning of contractile elements [18]. Therefore, the main difference between the two would be the mechanism of action, as DN provokes a mechanical disruption, whereas BTX-A works via a chemical denervation [17]. DN is considered to be an effective and safe treatment to improve function and spasticity in stroke patients [13,14,19] when applied by an experienced physiotherapist [20]. Moreover, although DN may have some adverse effects—such as bruising, bleeding and pain—it does not have the other adverse effects that BTX-A can have, such as weakness, anatomic denervation, or long-term immune resistance [21]. However, when compared with BTX-A, DN has fewer long-lasting effects, which would involve including more treatment sessions [13].

To the authors' knowledge, there has only been one study analyzing the cost-effectiveness of DN in neurological conditions [22]—specifically in stroke patients in the subacute phase—showing that the addition of four sessions of DN treatment for the upper extremity appears to be cost-effective. Cost-effectiveness analysis of DN treatment is necessary in order to determine the economic impact of adding or substituting treatments in the routine clinical practice of healthcare centers. Therefore, the main aim of this study was to analyze the cost-effectiveness of DN in patients with stroke in a chronic phase, using a cost–utility analysis in EUR per quality-adjusted life years (QALY), as well as analyzing the response to treatment based on the Modified Modified Ashworth Scale (MMAS) as the main outcome variable.

2. Materials and Methods

2.1. Study Design

An economic evaluation was performed following a previous randomized controlled trial (RCT) conducted at the Aragon Association of Stroke in Zaragoza (Spain). The study was approved by the Ethics Committee of Aragon (reference no. P116/0160), and registered at Clinicaltrials.gov no. NCT03546517 on June 2018.

2.2. Participants

The sample size estimation in the RCT was 23 participants [14]. Patients were enrolled in the study if they fulfilled the following criteria: (1) 40–90 years old, with hemiparesis from stroke of more than six months evolution; (2) ability to follow instructions and reply to questionnaires; and (3) hypertonia in at least one of the muscles of the upper extremity according to the MMAS score. The exclusion criteria were as follows: (1) grade 0 or 4 of hypertonia measured with the MMAS; (2) previous treatment with BTX-A or other treatments for hypertonia in the previous six months; (3) other neurodegenerative conditions; (4) fear of needles or contraindication to treatment with DN; and (5) cognitive decline ($\leq$24 points on the Mini-Mental State Examination test). All participants provided signed informed consent before participation in the study.

2.3. Intervention Conditions

There were two groups: the intervention group (IG), and the sham group (SG). Patients were randomized with a 1:1 allocation ratio using an online research randomizer sequence generator (http://www.randomizer.org (accessed on 9 June 2018)). After randomization, the physiotherapist performed the treatment according to the assignments. The IG received a single-session treatment of DN with the DNHS® technique, whereas the SG received a sham intervention, placing the needles superficially at the skin level and simulating the intervention in the same treatment context [14,23,24]. The DNHS® technique is an adaptation of traditional DN techniques, with specific diagnostic and application criteria for neurological patients. One of the main differences is that muscles are needled in submaximal stretch, and progression with the treatment is based on spasticity release and not on pain recognition [14,25]. The muscles treated were the biceps brachii and brachialis, flexor digitorum superficialis and profundus, extensor digitorum, adductor pollicis, and triceps brachii. Diagnosis of myofascial trigger points in the muscles selected was performed with clinical criteria, which have proven to be valid and reliable [26].

2.4. Main Measures

The two variables used for the cost-effectiveness study were the values of the EuroQol-5D questionnaire and the MMAS.

2.4.1. EuroQol-5D

The EQ-5D (EQ-5D-3L) is a recognized patient-reported outcome questionnaire including five dimensions (mobility, self-care, usual activities, pain/discomfort, and anxiety/depression) with three levels each (no problems, some problems, and extreme problems). The EQ-5D-3L questionnaire has a Cronbach's alpha value of 0.928, and it has proven to be a reliable tool to measure QOL among stoke survivors [27,28].

2.4.2. Modified Modified Ashworth Scale

The MMAS is a clinical scale used to assess hypertonia, and has been widely used despite its subjective component [29]; it scores the resistance to passive movement ranging from 0 (no increase in muscle tone) to 4 (rigidity in flexion or extension), and it has exhibited a good intra- (ICC = 0.748) and inter-rater (ICC = 0.781) reliability for assessing hypertonia in persons who have had a stroke [30]. Flexor and extensor muscles of the elbow and wrist are evaluated by assessing the resistance when the affected muscle group is passively stretched [31].

An improvement of at least one point in the MMAS is considered to be a clinically significant change [32]. The percentage of responders to treatment according to the MMAS has already been used in previous stroke cost-effectiveness studies in order to directly assess the impact of this type of intervention on affected muscles [22,32].

2.5. Costs

The evaluation of costs was carried out from the perspective of the hospital, clinic, or health center. From this point of view, only the direct healthcare costs associated with the DNHS® technique were considered: the DN materials and the cost of the physiotherapist session. The materials used were gauze, disinfectant, and needles, and the cost of the session was determined based on public data from the official bulletins of five representative regions of Spain (Aragón [33], Castilla y León [34], Madrid [35], País Vasco [36], and Cataluña [37]). These bulletins publish the prices established for the provision of health services outside the public health system.

2.6. Outcomes

2.6.1. Quality of Life (QOL)

QOL was measured at the beginning and two weeks after the DN intervention, and then the QALY was estimated using the area under the curve analysis. QALY is the preferred measure of health outcomes for use in technological appraisals, because it combines the impact of gains in QOL and in quantity of life (years) associated with an intervention [38]. The economic analysis through QOL was carried out by combining these data with the costs of the two interventions (DN and sham-DN), determining the incremental cost-effectiveness ratio (ICER). The ICER was calculated by dividing the difference in total costs by the difference in QOL, which represents the extra cost per extra unite of QOL [39]. The data were aggregated to the cost-effectiveness plane (graph), and were compared with the accepted cost-effectiveness threshold of EUR 20,000/QALY [40].

2.6.2. Modified Modified Ashworth Scale (MMAS)

MMAS values were obtained before and after the first DN session, and again two weeks later. In order to be able to use these values in the economic analysis, it was necessary to transform this information into the number of responders to treatment. The number of responders was determined for each muscle, and the data were related to the costs of the intervention, indicating the cost per responder to treatment and the incremental cost-effectiveness ratio.

2.7. Statistical Analysis

Data were analyzed with SPSS version 25.0 (SPSS, Inc., Chicago, IL, USA), and were plotted with Excel (Microsoft, Redmond, WA, USA). Significant differences between quality-of-life measurements in the RCT [14] were confirmed, and the information was completed by McNemar's test in the case of MMAS responders. To complete the sensitivity of the study, the minimum and maximum QOL values were added, and the data were compared to a cost-effectiveness threshold on a cost-effectiveness plane.

3. Results

A total of 23 patients aged 60.87 ± 15.16 years (mean $\pm$ SD; 61% male) were included in the final economic analysis; 11 were allocated to the IG and 12 to the SG, with no statistically significant differences between the groups at baseline.

3.1. Costs

The established costs of the treatment are displayed in Table 1. The DN materials used for one session cost EUR 0.64, and the mean physiotherapy cost per session according to published bulletins is EUR 14.32 ± 4.39. The DN material cost was the only difference between the IG and SG.

Table 1. Cost of treatment.

	Unitary Cost	Sham Group	Intervention Group
Dry needling (material per session)	EUR 0.64	-	EUR 0.64
Mean physiotherapy cost per session	EUR 14.32 ± 4.39	EUR 14.32	EUR 14.96

3.2. Quality of Life

The RCT results showed significant differences between groups in terms of QOL two weeks after the intervention (Table 2). The resulting QALYs were higher in the IG (Figure 1).

Table 2. QOL, QALY, and ICER.

		Sham Group	Intervention Group
QOL	Pre-test	0.01 ± 0.16	0.09 ± 0.43 *
	2 weeks	0.005 ± 0.06	0.18 ± 0.47 *
QALY		0.0003 (min. −0.0077; max. 0.0115)	0.0052 (min. −0.0230, max. 0.02493)
ICER (EUR/QALY)		130.14 (min. −41.57, max. 47.51)	

* $p < 0.05$ within IG and between IG and SG.

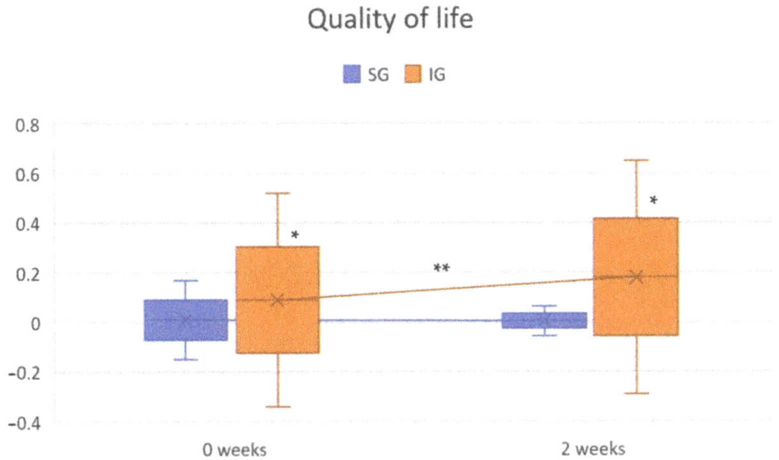

Figure 1. Variation of QOL during the study timeline. Abbreviations—IG: intervention group; SG: sham group. * $p < 0.05$ within IG; ** $p < 0.05$ between IG and SG.

Costs and QALY are represented in a cost-effectiveness plane, with the accepted cost-effectiveness threshold of EUR 20,000 in red.

As can be seen in Figure 2, the values of IG, SGmax, and IGmax are below the cost-effectiveness threshold, which corresponds to the most favorable values for accepting the inclusion of this treatment. However, the values of SG, IGmin, and SGmin imply higher costs and/or lower QALY values.

3.3. Modified Modified Ashworth Scale

An improvement of one point on this scale in one of the five movements assessed was considered a response to treatment. In Table 3, it can be seen that the percentage of responders in the IG was higher in all of the muscles except for the elbow flexors. From the percentage of responders, it can be seen that the post-intervention cost per responder in the IG is on average almost half that of the control, and that at two weeks, the cost per responder remains the same in the IG, but the cost drops in the control (SG). However, the results of this scale only showed statistically significant improvements in the elbow extensors, with 73% of patients responding in the IG vs. 8% responding in the SG, considering the values taken before and just after the session with DN. The resulting ICER was low, indicating that DN only costs EUR 0.99 more than the alternative without DN to get one more respondent. The mean rate of treatment responders was higher in the IG than in the SG (39% vs. 20% after session and 41% vs. 28% two weeks after the treatment).

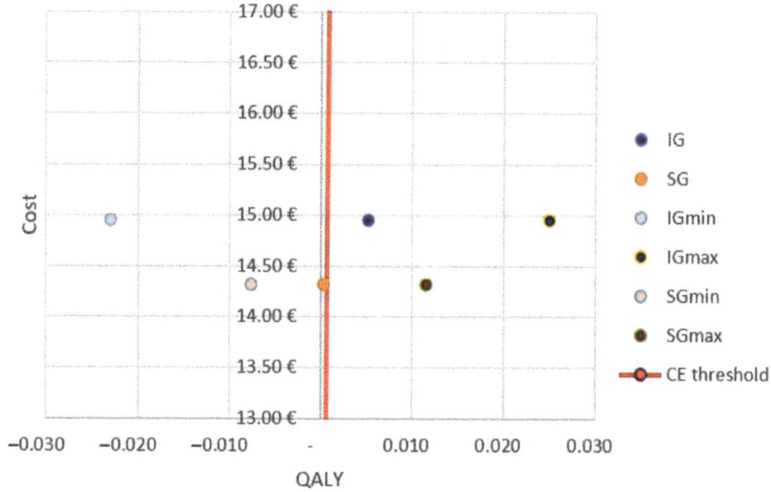

Figure 2. Cost-effectiveness plane. The average, minimum, and maximum values of the IG and SG are shown on the right-hand side of the cost-effectiveness threshold of EUR 20,000/QALY. Abbreviations—IG: intervention group; SG: sham group.

Table 3. Rate of responders to treatment and cost per responder.

	Post-Intervention									2 Weeks								
	Control			Intervention (DNHS®)						Control			Intervention (DNHS®)					
	n	% Responder	EUR/ Responder	n	% Responder	EUR/ Responder	McNemar Test	ICER		n	% Responder	EUR/ Responder	n	% Responder	EUR/ Responder	McNemar Test	ICER	
Elbow flexors	12	33%	EUR 42.96	11	27%	EUR 54.85	1.000	EUR −10.52		12	33%	EUR 42.96	11	27%	EUR 54.85	1.000	EUR −10.52	
Elbow extensors	12	8%	EUR 171.84	11	73%	EUR 20.57	0.039 *	EUR 0.99		12	25%	EUR 57.28	11	73%	EUR 20.57	0.125	EUR 1.34	
Wrist–dorsal flexors	12	25%	EUR 57.28	11	36%	EUR 41.13	1.000	EUR 5.61		12	17%	EUR 85.92	11	45%	EUR 32.91	0.375	EUR 2.22	
Wrist–palmar flexors	12	17%	EUR 85.92	11	18%	EUR 82.27	1.000	EUR 42.09		12	17%	EUR 85.92	11	27%	EUR 54.85	1.000	EUR 6.01	
Thumb adductor	12	17%	EUR 85.92	10	40%	EUR 37.39	0.375	EUR 2.73		12	50%	EUR 28.64	10	30%	EUR 49.86	0.625	EUR −3.19	
Means		20%	EUR 88.78		39%	EUR 47.24		EUR 8.18			28%	EUR 60.14		41%	EUR 42.60		EUR −0.83	

* $p < 0.05$ for elbow extensors post-intervention.

4. Discussion

This study explored the relationships between costs, QOL, and hypertonia when DN is applied to the upper extremity muscles of persons with chronic stroke, with the aim of analyzing the cost-effectiveness of a single DN session via a cost–utility analysis. Regarding costs, the main contributors when applying DN are the staff costs, since the materials used for DN have a very low cost. Although there are no fixed prices for the application of DN, and it may vary from country to country, in the case of Spain it can be calculated with relative certainty thanks to the official bulletins that publish the prices paid for services outside the Spanish public health system, as has been shown in previous studies [22,41].

Subsequently, the two variables we chose to assess the effectiveness of treatment were the QOL and the treatment response rate. The RCT results showed an improvement in QOL two weeks after the DN intervention, with significant differences between groups [14]. Other studies have also used the EQ-5D to assess the effectiveness of treatments in stroke patients but, unlike this study, they found no significant differences in quality of life between groups when applying DN [13] or BTX-A treatment [42]. It is possible that the limited number of patients included in the study may have influenced this, as the results were clearly better in the IG, with +0.0049 QALY compared to the SG.

On the other hand, we have a clinical variable, MMAS, which allows us to easily determine which patients are responders to treatment from a perspective other than that of the EQ-5D. In this case, significant differences were only obtained before and after treatment in elbow extensors, where we observed 65% more responders in the IG than in the SG. We remain unsure of the reasons for such a variety of responses to treatment, as well as why some muscle groups respond to treatment very differently than others, which is something worthy of consideration in future cost-effectiveness studies.

The results of the cost-effectiveness analysis indicate very low ICER values for both QOL and responder rate. The representation in Figure 2 allows us to see that the IG alternative is placed in the quadrant with the best cost-effectiveness ratio. However, the sensitivity analysis performed with the maximum and minimum values does not allow us to confirm the dominance (higher effectiveness with less cost) of the IG over the SG, since the minimum values of the IG could present worse QOL data, and the maximum values obtained for QOL in the SG show a cost-effective balance. In the economic analysis with the MMAS, the cost per respondent is EUR 41.54 less if valued just after the session, or EUR 17.54 less if valued two weeks after treatment. These results are consistent with another previously published study [22], and might suggest that adding DN to upper extremity rehabilitation treatment in stroke patients is a good alternative. The main reason for this is the low cost of this type of intervention which, combined with even a slight improvement in effectiveness variables, gives favorable results in the ICER. According to our data, to obtain one more QALY than the control group, we would only need to invest an additional EUR 130.14, and for responder patients the cost would be EUR 8.18 after treatment, or even a saving of EUR 0.83 at 2 weeks.

This study has certain limitations, such as the low number of patients evaluated, short follow-up period, and the use of only a single DN session. It is possible that a higher number of sessions may have a greater influence on QOL, as seen in the study of Cuenca Zaldívar et al., 2021 [13], where six needling sessions were performed. Moreover, the performance of only a single DN session limits the possibility of giving recommendations for the optimal number of sessions in terms of cost-effectiveness for the treatment of people with chronic stroke, which is something that could be done in the study carried out by Cuenca Zaldívar et al., who concluded that in the case of people with subacute stroke, four sessions of DN were more cost-effective than six sessions [13].

Since the first publication of the application of DN for spasticity in a child with cerebral palsy in 2007 [43], the number of studies of DN has increased and extended to many different neurological conditions, such as stroke [14], Parkinson's disease [15], and multiple sclerosis [16]. However, although recent systematic reviews and meta-analyses have shown that DN is effective in decreasing spasticity, there has only been one study of cost-effectiveness in subacute stroke patients and, therefore, this study contributes to increase the evidence for the inclusion of DN treatment in rehabilitation protocols from the clinical and healthcare perspectives. This is important considering not only the effectiveness, but also the low costs compared to botulinum toxin, as the annual cost of the toxin vials for adult patients with spasticity in the upper extremity can range between EUR 529.87 and 1180.72 [44]. Moreover, from the patient's perspective, it is important to consider that the treatment for focal spasticity is limited to BTX-A infiltrations [45] and, therefore, researching new treatment alternatives allows patients to empower themselves and seek treatments that are evidence-based, but under a patient-centered approach, considering their preferences [46,47].

Future lines of research should perform secondary analysis of RCTs of DN and other non-pharmacological interventions, including costs and cost-effectiveness. Apart from economic reasons derived from the costs of some pharmacological interventions, such as BTX-A infiltration, non-pharmacological treatments also have fewer adverse effects, and more research should be carried out from both the clinical and healthcare system perspectives, in order to analyze whether they can constitute an alternative to pharmacological treatments.

Although these results should be considered with caution due to the aforementioned limitations, the analysis performed shows that the inclusion of a single session of DN in the upper extremity rehabilitation protocols for chronic stroke patients can be of great benefit; therefore, future studies with a longer follow-up and a larger number of patients should be conducted in order to confirm these findings.

5. Conclusions

In this study, a cost-effectiveness analysis was conducted using two different effectiveness outcomes: the EQ-5D for QOL, and the treatment responders according to their hypertonia measured using the MMAS. The findings regarding the rate of responders showed good results in the cost-effectiveness analysis after treatment and at two weeks follow-up, finding that the application of DN in the upper extremity is an affordable alternative to use in patients with chronic stroke.

Author Contributions: Conceptualization: D.F.-S., S.C., and P.H.; methodology, D.F.-S. and P.H.; investigation, S.C., C.J.-S., N.B.-d.l.C., and P.H.; writing—original draft preparation, D.F.-S., N.B.-d.l.C., C.J.-S., M.G.-C., S.C., and P.H.; writing—review and editing, D.F.-S., N.B.-d.l.C., C.J.-S., M.G.-C., S.C., and P.H.; supervision, P.H. All authors have read and agreed to the published version of the manuscript.

Funding: This research received no external funding.

Institutional Review Board Statement: The clinical study associated with this secondary analysis was conducted according to the guidelines of the Declaration of Helsinki and approved by the Ethics Committee of Aragon (reference no. PI16/0160).

Informed Consent Statement: Informed consent to publish this paper was obtained from all subjects involved in the study.

Data Availability Statement: Data are available on reasonable request by email to efernandez@usj.es.

Acknowledgments: The authors acknowledge the participation of the members and staff of the Aragon Association of Stroke in Zaragoza (Spain).

Conflicts of Interest: The authors declare the following potential conflicts of interest with respect to the research, authorship, and/or publication of this article: The DNHS® technique was registered by Pablo Herrero. The other authors of this work declare that there are no potential conflicts of interest regarding the research, authorship, and/or publication of this article.

References

1. National Statistics Institute; Spanish Statistical Office. INEbase. Instituto Nacional de Estadística (INE). Available online: https://www.ine.es/dyngs/INEbase/es/operacion.htm?c=Estadistica_C&cid=1254736176780&menu=ultiDatos&idp=1254735573175 (accessed on 20 May 2021).
2. Kuriakose, D.; Xiao, Z. Pathophysiology and Treatment of Stroke: Present Status and Future Perspectives. *Int. J. Mol. Sci.* **2020**, *21*, 7609. [CrossRef] [PubMed]
3. Ma, V.Y.; Chan, L.; Carruthers, K.J. Incidence, Prevalence, Costs, and Impact on Disability of Common Conditions Requiring Rehabilitation in the United States: Stroke, Spinal Cord Injury, Traumatic Brain Injury, Multiple Sclerosis, Osteoarthritis, Rheumatoid Arthritis, Limb Loss, and Back Pain. *Arch. Phys. Med. Rehabil.* **2014**, *95*, 986–995.e1. [CrossRef] [PubMed]
4. Pan, F.; Hernández, L.; Ward, A. Cost-effectiveness of stroke treatments and secondary preventions. *Expert Opin. Pharmacother.* **2012**, *13*, 1751–1760. [CrossRef] [PubMed]
5. Thibaut, A.; Chatelle, C.; Ziegler, E.; Bruno, M.-A.; Laureys, S.; Gosseries, O. Spasticity after stroke: Physiology, assessment and treatment. *Brain Inj.* **2013**, *27*, 1093–1105. [CrossRef] [PubMed]
6. Schinwelski, M.J.; Sitek, E.; Waz, P.; Slawek, J. Prevalence and predictors of post-stroke spasticity and its impact on daily living and quality of life. *Neurol. i Neurochir. Polska* **2019**, *53*, 449–457. [CrossRef] [PubMed]
7. Moggio, L.; De Sire, A.; Marotta, N.; Demeco, A.; Ammendolia, A. Exoskeleton versus end-effector robot-assisted therapy for finger-hand motor recovery in stroke survivors: Systematic review and meta-analysis. *Top. Stroke Rehabil.* **2021**, 1–12. [CrossRef]
8. Lee, H.S.; Park, Y.J.; Park, S.W. The Effects of Virtual Reality Training on Function in Chronic Stroke Patients: A Systematic Review and Meta-Analysis. *BioMed Res. Int.* **2019**, *2019*, 7595639. [CrossRef]
9. Veerbeek, J.M.; Van Wegen, E.; Van Peppen, R.; Van Der Wees, P.J.; Hendriks, E.; Rietberg, M.; Kwakkel, G. What Is the Evidence for Physical Therapy Poststroke? A Systematic Review and Meta-Analysis. *PLoS ONE* **2014**, *9*, e87987. [CrossRef]
10. Kinnear, B.Z.; Lannin, N.A.; Cusick, A.; A Harvey, L.; Rawicki, B. Rehabilitation Therapies After Botulinum Toxin-A Injection to Manage Limb Spasticity: A Systematic Review. *Phys. Ther.* **2014**, *94*, 1569–1581. [CrossRef]
11. Sánchez-Mila, Z.; Salom-Moreno, J.; Fernández-De-Las-Peñas, C. Effects of Dry Needling on Post-Stroke Spasticity, Motor Function and Stability Limits: A Randomised Clinical Trial. *Acupunct. Med.* **2018**, *36*, 358–366. [CrossRef]
12. Ghannadi, S.; Shariat, A.; Ansari, N.N.; Tavakol, Z.; Honarpishe, R.; Dommerholt, J.; Noormohammadpour, P.; Ingle, L. The Effect of Dry Needling on Lower Limb Dysfunction in Poststroke Survivors. *J. Stroke Cerebrovasc. Dis. Off. J. Natl. Stroke Assoc.* **2020**, *29*, 104814. [CrossRef]
13. Zaldívar, J.N.C.; Calvo, S.; Bravo-Esteban, E.; Ruiz, P.O.; Santi-Cano, M.J.; Herrero, P. Effectiveness of dry needling for upper extremity spasticity, quality of life and function in subacute phase stroke patients. *Acupunct. Med.* **2021**, *39*, 299–308. [CrossRef]
14. Calvo, S.; La Cruz, N.B.-D.; Jiménez-Sánchez, C.; Bravo-Esteban, E.; Herrero, P. Effects of dry needling on function, hypertonia and quality of life in chronic stroke: A randomized clinical trial. *Acupunct. Med.* **2021**. [CrossRef]
15. La Cruz, N.B.-D.; Calvo, S.; Rodríguez-Blanco, C.; Herrero, P.; Bravo-Esteban, E. Effects of dry needling on gait and muscle tone in Parkinson's disease: A randomized clinical trial. *Acupunct. Med.* **2022**, *40*, 3–12. [CrossRef]
16. Luque-Moreno, C.; Granja-Domínguez, A.; Moral-Munoz, J.; Izquierdo-Ayuso, G.; Lucena-Anton, D.; Heredia-Rizo, A. Effectiveness of Dry Needling versus Placebo on Gait Performance, Spasticity, Electromyographic Activity, Pain, Range-of-Movement and Quality of Life in Patients with Multiple Sclerosis: A Randomized Controlled Trial Protocol. *Brain Sci.* **2020**, *10*, 997. [CrossRef]
17. Dong, Y.; Wu, T.; Hu, X.; Wang, T. Efficacy and safety of botulinum toxin type A for upper limb spasticity after stroke or traumatic brain injury: A systematic review with meta-analysis and trial sequential analysis. *Eur. J. Phys. Rehabil. Med.* **2017**, *53*, 256–267. [CrossRef]
18. Kütük, S.G.; Özkan, Y.; Kütük, M.; Özdaş, T. Comparison of the Efficacies of Dry Needling and Botox Methods in the Treatment of Myofascial Pain Syndrome Affecting the Temporomandibular Joint. *J. Craniofacial Surg.* **2019**, *30*, 1556–1559. [CrossRef]
19. Valencia-Chulián, R.; Heredia-Rizo, A.M.; Moral-Munoz, J.A.; Lucena-Anton, .D.; Luque-Moreno, C. Dry needling for the management of spasticity, pain, and range of movement in adults after stroke: A systematic review. *Complement. Ther. Med.* **2020**, *52*, 102515. [CrossRef]
20. Brady, S.; McEvoy, J.; Dommerholt, J.; Doody, C. Adverse events following trigger point dry needling: A prospective survey of chartered physiotherapists. *J. Man. Manip. Ther.* **2013**, *22*, 134–140. [CrossRef]
21. Kaku, M.; Simpson, D. Spotlight on botulinum toxin and its potential in the treatment of stroke-related spasticity. *Drug Des. Dev. Ther.* **2016**, *10*, 1085–1099. [CrossRef]
22. Sanchis, D.F.; Zaldívar, J.N.C.; Calvo, S.; Herrero, P.; Barrera, M.G. Cost-effectiveness of upper extremity dry needling in the rehabilitation of patients with stroke. *Acupunct. Med.* **2021**. [CrossRef]
23. Mitchell, U.H.; Stoneman, P.; Larson, R.E.; Page, G.L. The Construction of Sham Dry Needles and Their Validity. *Evid. Based Complement. Altern. Med.* **2018**, *2018*, 9567061. [CrossRef]
24. Braithwaite, F.A.; Walters, J.L.; Moseley, G.L.; Williams, M.T.; McEvoy, M.P. Towards more homogenous and rigorous methods in sham-controlled dry needling trials: Two Delphi surveys. *Physiotherapy* **2020**, *106*, 12–23. [CrossRef]
25. Herrero, P.; Calvo, S.; Ortiz, M. Dry Needling for Hypertonia an Spasticity (DNHS). In *Advanced Techniques in Musculoskeletal Medicine & Physiotherapy Using Minimally Invasive Therapies*; Minaya, I.F.V.F., Ed.; Chruchill-Livingston; Elsevier: London, UK, 2015.

26. Del Moral, O.M.; Lacomba, M.T.; Russell, I.J.; Méndez, Ó.S.; Sánchez-Sánchez, B. Validity and Reliability of Clinical Examination in the Diagnosis of Myofascial Pain Syndrome and Myofascial Trigger Points in Upper Quarter Muscles. *Pain Med.* **2018**, *19*, 2039–2050. [CrossRef]
27. Mahesh, P.K.B.; Gunathunga, M.W.; Jayasinghe, S.; Arnold, S.M.; Senanayake, S.; Senanayake, C.H.; De Silva, L.S.D.; Kularatna, S. Construct validity and reliability of EQ-5D-3L for stroke survivors in a lower middle income setting. *Ceylon. Med. J.* **2019**, *64*, 52–58. [CrossRef]
28. Pinto, E.; Maso, I.; Vilela, R.N.R.; Santos, L.C.; Oliveira-Filho, J. Validation of the EuroQol quality of life questionnaire on stroke victims. *Arq. Neuro-Psiquiatr.* **2011**, *69*, 320–323. [CrossRef]
29. Bohannon, R.W.; Smith, M.B. Interrater Reliability of a Modified Ashworth Scale of Muscle Spasticity. *Phys. Ther.* **1987**, *67*, 206–207. [CrossRef]
30. Meseguer-Henarejos, A.-B.; Sánchez-Meca, J.; López-Pina, J.-A.; Carles-Hernández, R. Inter- and intra-rater reliability of the Modified Ashworth Scale: A systematic review and meta-analysis. *Eur. J. Phys. Rehabil. Med.* **2018**, *54*, 576–590. [CrossRef]
31. Ansari, N.N.; Naghdi, S.; Mashayekhi, M.; Hasson, S.; Fakhari, Z.; Jalaie, S. Intra-rater reliability of the Modified Modified Ashworth Scale (MMAS) in the assessment of upper-limb muscle spasticity. *Neurorehabilitation* **2012**, *31*, 215–222. [CrossRef]
32. Rychlik, R.; Kreimendahl, F.; Schnur, N.; Lambert-Baumann, J.; Dressler, D. Quality of life and costs of spasticity treatment in German stroke patients. *Health Econ. Rev.* **2016**, *6*, 27. [CrossRef]
33. BOA ORDEN SAN/1221/2017, de 21 de Julio, Por la Que se Establecen los Precios y Tarifas Máximas Aplicables en la Prestación de Servicios Sanitarios con Medios Ajenos al Sistema de Salud de Aragón. Available online: http://www.boa.aragon.es/cgi-bin/EBOA/BRSCGI?CMD=VEROBJ&MLKOB=977342223030 (accessed on 22 October 2021).
34. Junta de Castilla y León. JRESOLUCIÓN de 21 de Noviembre de 2011, del PRESIDENTE de la Gerencia Regional de Salud, por la Que se Fijan las Tarifas Máximas y los Porcentajes de Revisión de las Condiciones Económicas Aplicables en el año 2011, a la Prestación de Servicios de Asistencia Sanitaria Concertada en el Ámbito de la Gerencia Regional de Salud. Available online: https://www.saludcastillayleon.es/institucion/es/resumen-bocyl-legislacion-sanitaria/resolucion-21-noviembre-2011-presidente-gerencia-regional-1 (accessed on 22 October 2021).
35. BOCM-20170821-1 ORDEN 727/2017, de 7 de agosto, del Consejero de Sanidad, por la que Se fijan los Precios Públicos por la Prestación de los Servicios y Actividades de Naturaleza Sanitaria de la Red de Centros de la Comunidad de Madrid. Available online: http://www.madrid.org/wleg_pub/secure/normativas/contenidoNormativa.jsf?opcion=VerHtml&nmnorma=9930#no-back-button (accessed on 22 October 2021).
36. TARIFAS PARA FACTURACIÓN DE SERVICIOS SANITARIOS Y DOCENTES DE OSAKIDETZA PARA EL AÑO 2019. Available online: https://www.osakidetza.euskadi.eus/contenidos/informacion/osk_servic_para_empresas/es_def/adjuntos/LIBRO-DE-TARIFAS_2020_osakidetza.pdf (accessed on 22 October 2021).
37. ORDRE SLT/42/2012, de 24 de Febrer, per la qual es Regulen els Supòsits i Conceptes Facturables i S'aproven els Preus Públics Corresponents als Serveis que Presta L'Institut Català de la Salut. Mar 2, 2012. Available online: http://ics.gencat.cat/web/.content/documents/transparencia/economia/ICF_2n2014.pdf (accessed on 22 October 2021).
38. York, Consortium YHE. Quality-Adjusted Life Year (QALY) 2016. Available online: https://yhec.co.uk/glossary/quality-adjusted-life-year-qaly/ (accessed on 22 October 2021).
39. YHE Consortium. Incremental Cost-Effectiveness Ratio (ICER). Available online: https://yhec.co.uk/glossary/incremental-cost-effectiveness-ratio-icer/ (accessed on 22 October 2021).
40. Vallejo-Torres, L.; García-Lorenzo, B.; Serrano-Aguilar, P. Estimating a cost-effectiveness threshold for the Spanish NHS. *Health Econ.* **2018**, *27*, 746–761. [CrossRef] [PubMed]
41. Arias-Buría, J.L.; Martín-Saborido, C.; Cleland, J.; Koppenhaver, S.L.; Plaza-Manzano, G.; Fernández-De-Las-Peñas, C. Cost-effectiveness Evaluation of the Inclusion of Dry Needling into an Exercise Program for Subacromial Pain Syndrome: Evidence from a Randomized Clinical Trial. *Pain Med.* **2018**, *19*, 2336–2347. [CrossRef] [PubMed]
42. Shaw, L.; Rodgers, H.; Price, C.; Van Wijck, F.; Shackley, P.; Steen, N.; Barnes, M.; Ford, G.; Graham, L. BoTULS: A multicentre randomised controlled trial to evaluate the clinical effectiveness and cost-effectiveness of treating upper limb spasticity due to stroke with botulinum toxin type A. *Health Technol. Assess.* **2010**, *14*, 1–113. [CrossRef] [PubMed]
43. Herrero, P.; Del Moral, O.M. A Case Study Looking at the Effectiveness of Deep Dry Needling for the Management of Hypertonia. *J. Musculoskelet. Pain* **2007**, *15*, 55–60. [CrossRef]
44. Andrés-Nogales, F.d.; Morell, A.; Aracil, J.; Torres, C.; Oyagüez, I.; Casado, M.A. Análisis de costes del uso de toxina botulínica A en España. *J. Farm. Hosp.* **2014**, *38*, 193–201.
45. Wissel, J. [Multidisciplinary therapy for focal spasticity treatment]. *Nervenarzt* **2008**, *79* (Suppl. 1), 24–28.
46. Dörfler, E.; Kulnik, S.T. Despite communication and cognitive impairment—person-centred goal-setting after stroke: A qualitative study. *Disabil. Rehabil.* **2020**, *42*, 3628–3637. [CrossRef]
47. Tonelli, M.R.; Sullivan, M.D. Person-centred shared decision making. *J. Eval. Clin. Pract.* **2019**, *25*, 1057–1062. [CrossRef]

healthcare

MDPI

Brief Report

The Association between Mental Motor Imagery and Real Movement in Stroke

Ana Poveda-García [1,*], Carmen Moret-Tatay [2,3] and Miguel Gómez-Martínez [4,5]

1 Escuela de Doctorado, Universidad Católica de Valencia San Vicente Mártir, San Agustín 3, Esc. A, Entresuelo 1, 46002 València, Spain
2 Facultad de Psicología, Universidad Católica de Valencia San Vicente Mártir, Avenida de la Ilustración, Burjassot, 46100 Valencia, Spain; mariacarmen.moret@ucv.es
3 Dipartimento di Neuroscienze Salute Mentale e Organi di Senso, La Sapienza Università di Roma, 00185 Rome, Italy
4 Departamento de Terapia Ocupacional, Centro Superior de Estudios Universitarios La Salle, 28023 Madrid, Spain; miguel.gomez@lasallecampus.es
5 Occupational Thinks Research Group, Centro Superior de Estudios Universitarios La Salle, 28023 Madrid, Spain
* Correspondence: ana.poveda@mail.ucv.es

Abstract: Background: Stroke is the main cause of disability in adults; the most common and long-term sequela is upper-limb hemiparesis. Many studies support the idea that mental motor imagery, which is related to the visualization of movement patterns, activates the same areas of the cortex as if the movement occurred. Objectives: This study aims to examine the capacity to elaborate mental motor images, as well as its relationship to loss of movement in the upper limbs after a stroke. Method: An observational study, in a sample of 39 adults who suffered a stroke, was carried out. The upper limb movement and functionality, cognitive disorders, the ability to visualize mental images, and activities of daily living were examined. Results: The results depicted a statistically significant correlation between the ability to visualize upper limb mental motor images with movement, functionality, and strength. In addition, a correlation between visual–spatial skills and mental visualization of motor ability and upper limb movement was found. Conclusions: These results suggest that the rehabilitation approach focused on the improvement of mental motor imagery could be of interest for the upper limb rehabilitation of movement and functionality.

Keywords: mental motor imagery; neurorehabilitation; stroke

Citation: Poveda-García, A.; Moret-Tatay, C.; Gómez-Martínez, M. The Association between Mental Motor Imagery and Real Movement in Stroke. *Healthcare* **2021**, *9*, 1568. https://doi.org/10.3390/healthcare9111568

Academic Editors: Grigorios Nasios, Lambros Messinis, Efthimios Dardiotis and Markos Sgantzos

Received: 17 August 2021
Accepted: 11 November 2021
Published: 17 November 2021

1. Introduction

A stroke occurs when the blood supply to part of the brain is interrupted or reduced, preventing brain tissue from receiving oxygen and nutrients. Brain cells begin to die in minutes. There are two main causes of stroke: ischemic, which occurs when a vessel supplying blood to the brain is obstructed, and hemorrhagic, in which a weakened blood vessel supplying the brain bursts. The consequence of stroke in both cases is a neurological deficit derived from the fact that a part or area of the brain stops working properly. The neurological deficit is one of the main causes of disability worldwide. [1].

There are over 13.7 million new strokes of all types each year worldwide. Every year, over 116 million years of healthy life are lost due to stroke [2]. The predominant long-term disability, which usually features the worst prognosis, is upper limb (UL) movement [3,4]. Approximately only one-third of people affected by stroke achieve fully functional UL recovery [3,5].

Thus far, a wide range of strategies and devices have been developed for the purpose of promoting upper limb motor recovery after stroke by taking advantage of the brain's ability to reorganize its neural networks after injury. The functional organization of the motor system is modified by use, and it has been suggested that use-dependent plasticity may play a major role in the recovery of function after stroke [6–8].

This approach includes neuromodulation techniques, such as transcranial magnetic stimulation or transcranial direct current stimulation, or sensory transformation techniques, such as mental practice/mental imagery or mirror therapy, which enhance use-dependent plasticity. More precisely, it should be noted that motor imagery (MI) is the mental representation of an action without a physical movement or muscular activation.

Research supports the idea that motor imagery should, to some extent at least, involve the same neuronal substrate as an executed movement [9]. Indeed, several studies reported that greater activation is achieved in the supplementary motor area (SMA), the premotor cortex (PMC), and the primary motor cortex (M1) in subjects during both executed and imagined movement [10–14]. Some brain areas that are activated during motor imagery belong to the neural network known to be involved in the early stage of motor control (i.e., motor programming) [10].

This finding has opened many interesting lines of research, which can be described as follows: (i) knowledge of MI for the application of therapy [15–19]; (ii) evidence regarding the neurophysiological bases of MI [20–23]; (iii) knowledge regarding the relationship between MI and physical movements [24,25]; (iv) analysis and validation of assessment tools to measure the ability to visualize movements [26,27]. Thus, this research aimed to examine the relationship between MI and real UL movements in individuals affected by stroke. Considering that the visualization of a movement can activate neurons in the same motor areas, the purpose of this study was to address the following research question: what happens if we cannot create the mental MI, as in the case of stroke? Would the ability to move be affected? In other words, this work aimed to deepen the knowledge about the relationship between mental imagery and motor deficits after stroke.

Many studies support the idea of functional equivalence between motor images and motor execution [28–30]. Most of us can imagine moving our fingers typing on a computer, but apparently, we cannot imagine typing faster than we can actually move our fingers [31,32]. A study that analyses the functional equivalence between images and action concludes that if an action cannot be physically performed, it cannot be imagined with a high functional equivalence [33]. Most of these studies have been carried out with healthy participants, analyzing functional equivalence by checking the temporal coupling between motor visualization and their subsequent motor performance of a task [32,34,35]. The innovative approach of the current research is focused on investigating the lack of ability to visualize motor patterns, and how this is related to an actual loss of UL movement and functionality. To this end, it is expected that people who suffered a stroke and are not capable of visualizing movement experience a greater UL handicap, with poor fine motor skills, muscle weakness or spasticity, and loss of functionality and independence.

2. Materials and Methods

2.1. Ethics

This research was conducted in Los Madroños Hospital, a leading medical center based in Madrid (Spain), specializing in neurological pathologies and their neurorehabilitation. The research was approved by the Research Ethics Committee of the Higher Center for University Studies La Salle Madrid.

The participants received an informative form on the goals of the current research. Informed consent was compulsory. The participants were volunteers and were not coerced; they were free to leave at any moment.

2.2. Study Design

This study aimed to analyze and correlate UL movement scales, cognitive instruments, and evaluations of the capability to visualize MI. Additionally, activities of daily living (ADL) independence measures, along with information related to demographic characteristics such as months since the injury, were also considered in this research. A direct relationship between MI and movement was expected, since, according to the previous literature, patients with this profile are considered to have higher difficulty in performing movements if they do not visualize them.

An observational approach was chosen. Lastly, participants were divided into two groups of early post-stroke patients (up to 6 months after stroke) and late post-stroke phase (e.g., over 6 months) to examine motor imagery over time.

2.3. Data Analysis

Data analysis was conducted using statistical software, IBM SPSS Statistics version 20 (IBM, Armonk, NY, USA). A Bivariable analysis and Pearson correlation coefficient were employed. In addition, the participants were divided into two groups considering their ability to perform MI. Data normality was examined, and depending on their results, the non-parametric Mann–Whitney U or the parametric t-test were used to determine statistical significance. Lastly, a 2 (MI groups) x2 (time after injury groups) ANOVA was carried out on the scores of Mental Evoking Images, Movements and Activities Questionnaire. The statistical significance level was set at $p < 0.05$.

2.4. Participants

Inclusion criteria:

- People over 18 years old;
- Participants who have suffered an ischemic or hemorrhagic stroke, with no time limit since the injury;
- Medically stable participants who can attend therapy;
- Participants who are able to give informed consent.

Exclusion criteria:

- Participants with previous pathology in their upper extremity, traumatic, neurological, or any other type of pathology that may affect the results of the assessment;
- Participants who have reported some neurological alteration before the stroke;
- Severe aphasia, memory disorders, attention disorders, visual and communication disorders, or other neural symptoms, which may interfere with this study.

A final sample of 39 adults volunteered to participate in the study (11 females and 28 males). Participants' age ranged from 49 to 84 years old (M_{Age} = 66.15; SD = 9.88). All the participants had suffered a stroke (19 ischemic and 20 hemorrhagic) and had different hemiparesis levels and showed different half-body affected (22 participants with left half-body affected and 17 with right half-body); the neuroimaging results were very varied, and in some cases, the diagnosis was not very precise. The main areas of involvement were the left and right middle cerebral artery (23%), nucleus basalis (12%), and others (64%). Time since injury ranged from 1 month to 4 years since the stroke (M_{Months} = 7.87 SD = 9.20). None of them had a previous pathology in their upper extremity (traumatic, neurological, or any other type that may affect the results of the assessment). It should be also noted that all the participants were medically stable, so they could attend therapy.

2.5. Measures

For data collection, the participants were examined using several assessment methods and tests, which were conducted at the Los Madroños Hospital. The employed assessment tools are described in the following subsections.

2.5.1. Upper Extremity Evaluations

A total of five instruments to evaluate the most affected UL were used: (1) Fugl–Meyer assessment for the upper extremity (FMA-UE) [36], which is adapted and validated for the Spanish population. This is a stroke-specific performance-based impairment index that has been extensively tested and found to have excellent properties. This is a quantitative examination instrument with a maximum score of 66 points. It is designed to assess motor functioning, sensation, and joint functioning in subjects with post-stroke hemiplegia. It is applied for both clinical and research proposes to determine disease severity, describe motor recovery, and plan and assess treatment; (2) ABILHAND scale [37], which is also translated into Spanish. This is an appropriate outcome measure for assessing upper extremity performance in daily activities in subjects with stroke. It is an interview-based assessment in which the patient is asked to estimate the ease or difficulty of performing a list of activities when carried out without assistance. This can involve any strategy used to carry out the activity; it is a self-report and not a physical demonstration of the activity; (3) nine-hole peg test (9-HPT) [38], which is translated into Spanish. This is a standardized, quantitative assessment used to measure finger dexterity that requires participants to repeatedly place and then remove nine pegs into nine holes, one at a time, as quickly as possible; (4) the Jebsen hand function test (JHFT) [39], which is a translated tool into Spanish and a standardized evaluative measure of functional hand motor skills. JHFT consists of 7 items that measure fine motor skills, weighted functional tasks, and non-weighted functional tasks; (5) dynamometry test [40], for which a Deyard Tech digital dynamometer (range 5–90 kg) was used in order to assess the ULs grip strength. During the dynamometry test, the position of the participant was seated with his back resting on the backrest and their feet on the floor, the arm resting on a table in front of them, with the elbow flexed at 90° and the wrist in a neutral position. The assessment consisted of 2 tests where the subject pressed a manual dynamometer, once with each hand, and the best strength result of the 2 tests performed with each hand was taken.

2.5.2. Motor Imagery Ability

In mental health practice, different tools are usually used to evaluate the capability to perform MI such as the Imagery Use Questionnaire (IUQ) [41], Movement Imagery Questionnaire (MIQ) [42], Movement Imagery Questionnaire-Revised (MIQ-R) [43], Vividness of Movement Imagery Questionnaire (VMIQ) [44], and Kinesthetic and Visual Imagery Questionnaire (KVIQ) [27].

For the proposes of the current research, the Mental Evoking Images, Movements and Activities Questionnaire (CEMIMA) was used due to several reasons: First, CEMIMA is a psychometrically robust instrument that can be used to measure the ability to create visual and kinesthetic mental images of the UL. Second, CEMIMA can evaluate the capability to visualize the isolated UL movement. Third, CEMIMA allows the examiner to collect qualitative information about the MI visualization process, including the difficulties that the participants may have, such as interferences or disruptions during the MI visualization.

CEMIMA scale was developed and validated in the Spanish language. The process of data collection was divided into 5 different steps: steps 1 and 2 encompassed data collection on mental visualization of one of the hands; step 3 included data collection on movement visualization of selected hand; in steps 4 and 5, data on mental visualization of the selected hand during an activity (e.g., moving a teaspoon or throwing and catching a ball) were collected. The questionnaire had positive/negative scores. Positive scores were based on a Likert scale from 0 to 10 to study the visualization and kinesthetic capacities to perceive movement. Regarding the negative scores, for each interference (detected by the examiner or related by the participant), a negative point was given. The final questionnaire score was calculated by adding both positive scores (visualization and kinesthetic perception) and subtracting the negative scores (interferences) [26].

2.5.3. Cognitive Evaluations

Due to the complexity of elaborate mental images, it was necessary that participants also made cognitive evaluations to obtain a better understanding of some aspects that may affect the capability to visualize MI.

Attentional span, planning capacity, and visual–spatial (VS) working memory were evaluated to detect if there could be any relationship between the capacity to visualize the movement mentally and its impact on the affected UL movement. The tools employed are described as follows: (1) digit span test [45] is included in the Wechsler Adult Intelligence Scale (WAIS) and the Wechsler Memory Scales (WMS). Participants read a sequence of numbers and are asked to repeat the same sequence back to the examiner in order (forward span) or in reverse order (backward span). Forward span captures attention efficiency and capacity. Backward span is an executive task particularly dependent on working memory. The digit span subtest can be scored as one summary value (this is the score that is age normed and contributes to summary scores in the Wechsler tests) or separately for forward and backward performance; (2) block design test [45] is a subtest that is administered as part of several of the Wechsler Intelligence tests, including the Wechsler Preschool and Primary Scale of Intelligence (WPPSI, the Wechsler Intelligence Scale for Children-fourth edition [46]) and the Wechsler Adult Intelligence Scale-fourth edition [45]. It is the main measure of VS and organizational processing abilities, as well as nonverbal problem-solving skills. As this is a time-based task, it is also influenced by fine motor skills. The individual is presented with identical blocks with surfaces of solid red, surfaces of solid white, and surfaces that are half red and half white. Using an increasing number of these blocks, the individual is required to replicate a pattern that the test administrator presents to them—first as a physical model and then as a two-dimensional picture; (3) zoo map test [47] is adapted and validated in the Spanish population. The test belongs to the behavioral assessment of dysexecutive syndrome (BADS). It measures executive functions, in particular the ability to organize, plan, and solve problems in order to achieve an objective. This is a test of planning. It provides information about the participant's ability to plan a route to visit 6 of a possible 12 locations in a zoo, first in a demanding, open-ended situation where little external structure is provided and then in a situation that involves simply following a concrete, externally imposed strategy.

2.5.4. Independence of Basic and Instrumental Activities of Daily Living

(1) Barthel index (BI) is adapted and validated to the Spanish population. The Barthel index is a valid measure of disability [48] that assesses the autonomy of basic activities in daily life. Additionally, it is an important method to evaluate the capacity of participants to conduct 10 different ADLs, considered as basic ADLs, making it possible to obtain a quantitative estimation about their independence level. The values assigned to each activity are based on the time and quantity of physical assistance required by the participant; (2) the Lawton instrumental activities of daily living scale (IADL) is an appropriate instrument to assess independent living skills [49], which is adapted and validated in the Spanish language. The scale covers eight functional domains: using the telephone, shopping, food preparation, housekeeping, laundry, transport, medication, and finances. Competence is rated according to descriptions of the subject's level of involvement/ability in each activity.

3. Results

3.1. Comparison among Groups regarding Their Capacity to Visualize MI

After analyzing the obtained results, the participants were divided into two groups depending on their ability to visualize MI (Table 1): Group-1 (G1) with a poor ability to visualize and Group-2 (G2) with a better ability to visualize MI. The average value of CEMIMA test results was employed to distinguish participants. G1 had an average value of 31.73 (SD = 11.66) and G2 had an average value of 60.20 (SD = 7.17).

The differences were statistically significant ($p \leq 0.05$) for BI, block design, ABILHAND, FMA-UE and all its subtests (arm, wrist, hand, coordination, sensation, passive movement, and pain), 9-HPT, JHFT, and dynamometry (see Table 1).

3.2. Correlations between Mental Imagery and Motor Function Measures

Table 2 shows the correlations between MI and motor function measures. There is a highly significant correlation between the motor function measures ABILHAND, FMA-UE, 9-HPT, JHFT, and dynamometry and the MI evaluated through CEMIMA. There is a strong correlation between the motor function measures, ABILHAND, FMA-UE, 9-HPT, JHFT, and dynamometry and the MI evaluated through CEMIMA.

Table 1. Comparison among visualization of MI groups.

Evaluations	Group 1 Poor Ability to Visualize Group		Group 2 Better Ability to Visualize Group		Sig **
Variables	Average (SD *)	CI 95%	Average (SD *)	CI 95%	
Age	62.21 (7.73)	58.48–65.94	69.90 (10.41)	65.02–74.78	0.13 t
Months since injury	9.11 (10.94)	3.83–14.38	6.00 (7.26)	3.30–10.10	0.412 Z
CEMIMA	31.73 (11.66)	26.11–37.36	60.20 (7.17)	56.84–63.55	0.000 t
BI	53.16 (35.28)	36.15–70.16	82.00 (30.49)	67.73–96.27	0.005 Z
IADL	3.16 (2.98)	1.72–4.60	4.40 (2.60)	3.18–5.62	0.125 Z
ABILHAND	36.95 (59.07)	8.48–65.42	139.55 (43.71)	119.09–160.01	0.000 Z
FMA-UE	26.42 (20.55)	16.51–36.33	55.70 (15.74)	48.33–63.07	0.000 Z
FMA-UE Arm	13.84 (11.27)	8.41–19.27	30.15 (8.71)	26.07–34.23	0.000 Z
FMA-UE wrist	3.11 (3.97)	1.19–5.02	8.45 (2.92)	7.08–9.82	0.000 Z
FMA-UE hand	5.42 (5.43)	2.80–8.04	12.10 (3.89)	10.28–13.92	0.000 Z
FMA-UE sensitive	7.32 (4.04)	5.37–9.26	10.15 (2.88)	8.80–11.50	0.014 Z
FMA-UE coordination	4.00 (1.29)	3.3820134.62	4.95 (1.39)	4.30–5.60	0.023 Z
FMA-UE Pain	20.47 (4.33)	18.38–22.56	22.75 (2.59)	21.54–23.96	0.034 Z
FMA-UE Mvt. Pas	21.9 (2.29)	20.68–22.90	23.10 (1.61)	22.34–23.86	0.008 Z
JHFT	671.37 (313.63)	520.21–822.54	210.13 (236.65)	99.37–320.89	0.001 Z
9-HPT	102.74 (35.53)	85.62–119.87	61.41 (40.45)	42.47–80.34	0.005 Z
Dynamometry	4.25 (6.45)	1.14–7.36	17.32 (11.48)	11.54–22.18	0.000 t
Zoo map test	1.47 (1.32)	0.79–2.15	1.58 (1.64)	0.79–2.37	0.935 Z
Block design	7.42 (3.30)	5.83–9.01	9.55 (3.22)	8.04–11.06	0.049 t
Digit span	10.56 (3.11)	8.90–12.22	10.15 (2.34)	9.05–11.25	0.700 Z

* SD: standard deviation ** Sig: significance. Depending on the normality, the Mann–Whitney U or *t*-test was used to determine significance, with Student's *t*-test and a z for the Mann–Whitney U. CEMIMA: the mental evoking images, movements and activities questionnaire, FMA-UE: Fugl–Meyer assessment for the upper extremity, JHFT: the Jebsen hand function test; BI: Barthel index; IADL: Lawton instrumental activities of daily living scale; 9-HPT: nine-hole peg test.

Table 2. Pearson's correlation coefficient between MI and motor and cognitive function measures.

Variables	1	2	3	4	5	6	7	8	9	10	11	12	13
1.CEMIMA	–												
2.FMA-UE	0.74 **	–											
3. 9-HPT	−0.58 **	−0.84 **	–										
4.JHFT	−0.71 **	−0.94 **	0.87 **	–									
5. Dynamo	0.65 **	0.81 **	0.80 **	−0.77 **	–								
6. ABILHAND	0.77 **	0.94 **	0.80 **	−0.97 **	0.77 **	–							
7. Block D	0.54 **	0.43 **	−0.34 *	−0.41 *	0.37 *	0.43 **	–						
8. Digitos	−0.48	-0.90	0.83	0.59	−0.10	−0.11	0.36 *	–					
9. Zoo Map	0.18	0.28	−0.32 *	−0.24	0.33 *	0.18	0.20	0.32	–				
10. Age	0.25	0.23	−0.16	−0.34 *	0.12	0.34 *	−0.13	0.10	−0.13	–			
11. M.S.Inj	−0.35 *	−0.27	0.26	0.33 *	−0.25	0.34 *	−0.42 **	0.02	−0.02	−0.27	–		
12. IADL	0.34 *	0.55 **	−0.46 *	−041 *	0.45 *	0.41 **	0.16	−0.09	0.36 *	−0.35 *	0.28	–	
13. BI	0.47 **	0.62 **	−0.55 **	−0.48 **	0.56 **	0.52 **	0.17	−0.21	0.25	−0.46	0.18	0.81 **	–

Level of significance evaluations: * $p < 0.05$, ** $p < 0.001$. FMA-UE: Fugl–Meyer assessment for the upper extremity; JHFT: the Jebsen hand function test; 9-HPT: nine-hole peg test; Dynamo: dynamometry; Block D: block design; Zoo Map: zoo map test; M.S. inj: months since injury; IADL: the Lawton instrumental activities of daily living scale; BI: Barthel index.

3.3. *Correlations Cognitive Function Measures*

There is a statistically significant relationship between CEMIMA and block design; hence, there is a correlation between the VS function and the ability to visualize MI. Results obtained from other cognitive instruments did not depict higher correlations with the other cognitive tests—namely, digit span and zoo map tests (Table 2).

Regarding the correlations between cognitive functions and UL movement, it is important to mention there is a statistically significant correlation between the block design test and all the UL movement scales analyzed (ABILHAND, FMA-UE, 9-HPT, JHFT, and dynamometry), but no correlation is observed between digit span and zoo map tests and the UL scales, although zoo map test and 9-HPT are related (Table 2).

3.4. Time since Injury and Groups

As previously described, participants were divided into two groups of early post-stroke participants (up to 6 months after stroke) and late post-stroke participants (over 6 months); as such, two different profiles are expected, where participants with a longer period from injury have poorer performance on motor imagery.

The χ^2 was carried out across the time after injury groups and the evocation groups. No statistically significant differences were found across them ($p > 0.05$), indicating that groups were homogeneously distributed (Table 3).

Table 3. Contingency tables across evocation groups and early and late post-stroke period.

Participants		Time after Injury		Total
Group	**Counts**	**<6m**	**>6m**	
Lower	Count	10	9	19
	% within row	52.632 %	47.368 %	100 %
	% within column	40.000 %	64.286 %	48.718 %
	% of total	25.641 %	23.077 %	48.718 %
Higher	Count	15	5	20
	% within row	75 %	25 %	100 %
	% within column	60 %	35.714 %	51.282 %
	% of total	38.462 %	12.821 %	51.282 %
Total	Count	25	14	39
	% within row	64.103 %	35.897 %	100 %
	% within column	100 %	100 %	100 %
	% of total	64.103 %	35.897 %	100%

After checking the underlying assumptions (Levene's $p = 0.07$ and qqplots), the ANOVA in CEMIMA scores indicated that differences across groups were statistically significant: $F_{(1,35)} = 72.95$; MSE = 6020.65; $p < 0.01$; $\eta^2 = 0.638$. As expected, differences across time groups after injury also reached the statistical level: $F_{(1,35)} = 6.35$; MSE = 534.70; $p < 0.0$; $\eta^2 = 0.05$. No interactions were found across factors under study.

4. Discussion

MI analysis is a complex process full of difficulties related to this therapeutic method. Nevertheless, the question of how the visualization of MI can be a useful therapeutic tool is of interest for many fields. The literature seems to indicate that MI reproduces motor efference and might be limited by an individual's normal motor experience [50] This theory postulates that the motor schema involved in real activity is reinforced during the visualization of MI. Particularly, the motor skills are reinforced when those images involve the same motor schemas as the real movement [51].

In many studies related to the visualization of MI after a stroke [9,52,53], the quality of the visualization is considered a useful element in different treatment techniques. Conversely, the objective of our research was different—we aimed to examine whether the ability to visualize MI was related to the capacity to move the UL and understand which other variables could affect the quality of the visualization.

The main findings could be described as follows: First, a statistically significant correlation between MI and motor function measures was found. This result suggests a strong relationship between the ability to visualize MI with the UL movement and functionality after a stroke, supporting previous results in which functional equivalence between motor images and motor execution is claimed [28–30].

The relation between MI and movement is clear—participants were not able to perform the movements if they could not visualize them; as claimed by Olsson et al., "*If you cannot do it, you will not think it*" [33]. In several CEMIMA items, the participants were challenged to visualize a ball in their hand, and subsequently, they had to throw and catch the ball repeatedly. One result of interest in this test was related to how participants with the worst MI visualization ability had a bad movement performance. Surprisingly, we observed that some participants reported that during the visualization, they had dropped the ball or they had not been able to throw it.

Second, there is a statistically significant relationship between the block design test, which measures the VS working memory and UL evaluation instruments (FMA-UE, 9-HPT, JHFT, and dynamometry). Therefore, the participants with poorer VS skills had a greater alteration in their capacity to move the UL. Their hindrance to managing the MI was linked to the neural network disorganization, making it impossible for them to perform a UL movement correctly.

Third, there is a relationship between MI and cognitive function measures, especially VS working memory and their capability to visualize MI.

As a result of the two previous statements, it would be essential to determine whether the participants' inability to generate MI is related to an alteration in their cognitive capacity, or whether it is due to an alteration of the motor planning caused by the brain injury. Therefore, future lines of research should address whether poorer cognitive mechanisms to perform MI visualization are due to an alteration in their VS skills, or motor planning is damaged, and therefore, individuals cannot access it. Consequently, it seems important to emphasize two statistically significant correlations related to VS skills. On the one hand, VS with the MI visualization and, on the other hand, VS with the affected UL movement. These correlations allow us to conceptualize how motor and cognitive systems are strongly interrelated and, therefore, cannot be separated in neurorehabilitation.

The literature has addressed the relationship between cognition and movement [54–56]. An example of the importance of the movement mental representation and mental images manipulation occurs in the rehabilitation technique called graded motor imagery when training the perception of laterality. This training is based on judging, through a sequence of photographs, if the part of the body shown corresponds to the right or left half-body, where the participants have to perform an adequate mental rotation using the working memory process [57]. It is considered that these types of cognitive strategies, as well as others related to the involvement of cognitive functions in motor function, can be very useful in UL rehabilitation [58].

Lastly, two well-defined groups of participants were of interest according to their MI ability. G2 had a relatively normal ability to visualize MI, while, the other group had a greater impediment to achieving it (G1). The latter group, G1, had a lower ability to visualize MI, poorer VS, worst dexterity, less strength, and consequently a reduction in their UL mobility and functionality; additionally, they were more ADLs dependent. Future research should examine these profiles, which may set standardized criteria and approaches for the MI technique and would be very useful to the neurorehabilitation scientific community.

To conclude, it should be noted that the ability to visualize MI and cognitive functions such as VS and UL movement seem to be related. Thus, it is important to approach them all together during the treatments. This could be of interest for new rehabilitation techniques seeking to induce a cortical reorganization that would be associated with the motor relearning through the achieving and improvement in the motor function [59]. In summary, in these techniques, we can find motor imagery, action observation therapy, or mirror therapy. One should bear in mind that these therapies have a multitude of benefits: they are low cost because they do not require expensive tools such as robots; furthermore, they allow the development and use of emerging technologies such as non-immersive virtual reality (VR) applications (e.g., phones and tablets), which could be used as therapeutic supplements, particularly when physical movement is limited, enabling participants to continue their treatment in their home after finalizing the treatment session; our phones or tablets could simply be enough for allowing the subjects to continue their treatment in their home after finalizing the treatment session.

As a statistically significant relationship was found between the ability to visualize MI and the time elapsed since the stroke, it seems important to note that the longer it was since the stroke, the more likely it was to observe a lower capability to visualize MI. This could be related to the decrease in the cortical representation of the UL, probably caused by disuse, which, in turn, was generated by their incapacity to move. Likewise, time after injury is a question of interest from two aspects in applied research. On the one hand, research in this area might help to improve current theoretical models. On the other, one should bear in mind that for specific interventions to be effective, therapy must be initiated as soon as possible. Current results support the last statement, also suggesting a lack of interaction with evocation groups.

The main study limitations are related to the characteristics of the research design. It was not possible to establish control procedures for potential confounding variables to avoid certain biases in results. Future lines of research should include the results of other techniques such as neural electronic physiological exams. Moreover, and as is sometimes the case in studies with clinical samples, a small sample size was employed, which limited the generalization of results. On the other hand, the time span between the incidence of stroke and the assessment variation is a question that caused variation in the current results, as it was addressed in a dichotomous way. However, the study makes it possible to establish the possible relationship between the variables involved to conduct analytical studies, and it is expected this could be of interest for future lines of research that address these issues. Nevertheless, the results in this study seem to support previous studies in the field of MI research. Increasing the number of studies of motor imagery would shed light on the use of this technique as means of therapy.

5. Conclusions

We can conclude that the capacity to visualize mental images and the capacity to produce a movement in the UL after a stroke are related to each other. Additionally, we also found other correlations between VS skills and the capacity to visualize MI, as well as with the UL movement.

These results are of interest for neurorehabilitation proposes based on the improvement of MI, which could be beneficial for the UL movement and functionality rehabilitation. Including use-dependent plasticity therapies such as imagery motor, action observation therapy, or mirror therapy could increase the chances of success.

Lastly, the appearance and increased use of emerging ICT in rehabilitation bring forth new possibilities for therapists and their patients to implement activities that allow them to achieve their objectives.

Author Contributions: Conceptualization, A.P.-G., C.M.-T. and M.G.-M.; methodology, A.P.-G., M.G.-M., and C.M.-T.; software, A.P.-G. and M.G.-M.; validation, A.P.-G., M.G.-M. and C.M.-T.; formal analysis, A.P.-G., M.G.-M. and C.M.-T.; investigation, A.P.-G. and M.G.-M.; resources, A.P.-G.; data curation, A.P.-G. and M.G.-M.; writing—original draft preparation, A.P.-G., M.G.-M. and C.M.-T.; writing—review and editing, A.P.-G., M.G.-M. and C.M.-T.; visualization A.P.-G.; supervision, A.P.-G., C.M.-T. and M.G.-M.; project administration, A.P.-G., C.M.-T. and M.G.-M. All authors have read and agreed to the published version of the manuscript.

Funding: This research received no external funding.

Institutional Review Board Statement: The study was conducted according to the guidelines of the Declaration of Helsinki and approved by the Research Ethics Committee of the Higher Center for University Studies La Salle Madrid (CSEULS-PI-111/2016).

Informed Consent Statement: Informed consent was obtained from all subjects involved in the study.

Data Availability Statement: The data that support the findings of this study are available from the corresponding author on reasonable request.

Acknowledgments: We wish to thank the patients of the Hospital Los Madroños for their participation in the study and the Hospital Los Madroños and for supporting the research.

Conflicts of Interest: The authors declare no conflict of interest.

References

1. Coupland, A.P.; Thapar, A.; Qureshi, M.I.; Jenkins, H.; Davies, A.H. The definition of stroke. *J. R. Soc. Med.* **2017**, *110*, 9–12. [CrossRef]
2. Wilkins, E.; Wilson, L.; Wickramasinghe, K.; Bhatnagar, P.; Leal, J.; Luengo-Fernandez, R.; Burns, R.; Townsend, N.; Rayner, M. *European Cardiovascular Disease Statistics 2017*; European Heart Network: Brussels, Belgium, 2017.
3. Lee, S.; Bae, S.; Jeon, D.; Kim, K.Y. The effects of cognitive exercise therapy on chronic stroke patients' upper limb functions, activities of daily living and quality of life. *J. Phys. Ther. Sci.* **2015**, *27*, 2787–2791. [CrossRef]
4. Beebe, J.; Lang, C. Active Range of Motion Predicts Upper Extremity Function 3 Months After Stroke. *Stroke* **2009**, *40*, 1772–1779. [CrossRef]
5. Dur Mata, M.J.; Molleda Marzo, M.; García Almazán, C.; Mallol Badellino, J.; Calderon Padilla, V. Factores pronósticos en el ictus. De la fase aguda a los tres años. *Rehabilitacion* **2011**, *45*, 18–23. [CrossRef]
6. Classen, J.; Liepert, J.; Wise, S.P.; Hallett, M.; Cohen, L.G. Rapid Plasticity of Human Cortical Movement Representation Induced by Practice. *J. Neurophysiol.* **1998**, *79*, 1117–1123. [CrossRef]
7. Bütefisch, C.M.; Davis, B.C.; Wise, S.P.; Sawaki, L.; Kopylev, L.; Classen, J.; Cohen, L.G. Mechanisms of use-dependent plasticity in the human motor cortex. *Proc. Natl. Acad. Sci. USA* **2000**, *97*, 3661–3665. [CrossRef]
8. Koganemaru, S.; Mima, T.; Thabit, M.N.; Ikkaku, T.; Shimada, K.; Kanematsu, M.; Takahashi, K.; Fawi, G.; Takahashi, R.; Fukuyama, H.; et al. Recovery of upper-limb function due to enhanced use-dependent plasticity in chronic stroke patients. *Brain* **2010**, *133*, 3373–3384. [CrossRef]
9. Decety, J. The neurophysiological basis of motor imagery. *Behav. Brain Res.* **1996**, *77*, 45–52. [CrossRef]
10. Schieber, M.H. Dissociating motor cortex from the motor. *J. Physiol.* **2011**, *589*, 5613–5624. [CrossRef]
11. Loporto, M.; McAllister, C.; Williams, J.; Hardwick, R.; Holmes, P. Investigating central mechanisms underlying the effects of action observation and imagery through transcranial magnetic stimulation. *J. Mot. Behav.* **2011**, *43*, 361–373. [CrossRef]
12. Jacobson, E. Electrical measures of neuromuscular states during mental activities: VI. A note on mental a ctivities concerning an amputated limb. *Am. J. Physiol.* **1931**, *96*, 122–125. [CrossRef]
13. Lotze, M.; Montoya, P.; Erb, M.; Hülsmann, E.; Flor, H.; Klose, U.; Birbaumer, N.; Grodd, W. Activation of cortical and cerebellar motor areas during executed and imagined hand movements: An fMRI study. *J. Cogn. Neurosci.* **1999**, *11*, 491–501. [CrossRef]
14. Butler, A.J.; Page, S.J. Mental Practice With Motor Imagery: Evidence for Motor Recovery and Cortical Reorganization After Stroke. *Arch. Phys. Med. Rehabil.* **2006**, *87* (Suppl. 12), 2–11. [CrossRef]
15. Morioka, S.; Osumi, M.; Nishi, Y.; Ishigaki, T.; Ishibashi, R.; Sakauchi, T.; Takamura, Y.; Nobusako, S. Motor-imagery ability and function of hemiplegic upper limb in stroke patients. *Ann. Clin. Transl. Neurol.* **2019**, *6*, 596–604. [CrossRef]
16. López, N.D.; Monge Pereira, E.; Centeno, E.J.; Miangolarra Page, J.C. Motor imagery as a complementary technique for functional recovery after stroke: A systematic review. *Top. Stroke Rehabil.* **2019**, *26*, 576–587. [CrossRef]
17. Fernandez-Gomez, E.; Sanchez-Cabeza, A. Motor imagery: A systematic review of its effectiveness in the rehabilitation of the upper limb following a stroke. *Rev. Neurol.* **2018**, *66*, 137–146.
18. Machado, T.C.; Carregosa, A.A.; Santos, M.S.; Ribeiro NM da, S.; Melo, A. Efficacy of motor imagery additional to motor-based therapy in the recovery of motor function of the upper limb in post-stroke individuals: A systematic review. *Top. Stroke Rehabil.* **2019**, *26*, 548–553. [CrossRef]
19. Gowda, A.S.; Memon, A.N.; Bidika, E.; Salib, M.; Rallabhandi, B.; Fayyaz, H. Investigating the Viability of Motor Imagery as a Physical Rehabilitation Treatment for Patients With Stroke-Induced Motor Cortical Damage. *Cureus* **2021**, *13*, e14001. [CrossRef]
20. Szameitat, A.J.; Shen, S.; Conforto, A.; Sterr, A. Cortical activation during executed, imagined, observed, and passive wrist movements in healthy volunteers and stroke patients. *Neuroimage* **2012**, *62*, 266–280. [CrossRef]
21. Li, F.; Zhang, T.; Li, B.J.; Zhang, W.; Zhao, J.; Song, L.P. Motor imagery training induces changes in brain neural networks in stroke patients. *Neural. Regen. Res.* **2018**, *13*, 1771–1781.
22. Wang, X.; Wang, H.; Xiong, X.; Sun, C.; Zhu, B.; Xu, Y.; Fan, M.; Tong, S.; Sun, L.; Guo, X. Motor Imagery Training After Stroke Increases Slow-5 Oscillations and Functional Connectivity in the Ipsilesional Inferior Parietal Lobule. *Neurorehabil. Neural Repair* **2020**, *34*, 321–332. [CrossRef]
23. Hardwick, R.; Caspers, S.; Eickhoff, S.; Swinnen, S. Neural correlates of action: Comparing meta-analyses of imagery, observation, and execution. *Neurosci. Biobehav. Rev.* **2017**, *94*, 31–44. [CrossRef]
24. Höller, Y.; Bergmann, J.; Kronbichler, M.; Crone, J.S.; Schmid, E.V.; Thomschewski, A.; Butz, K.; Schütze, V.; Höller, P.; Trinka, E. Real movement vs. motor imagery in healthy subjects. *Int. J. Psychophysiol.* **2013**, *87*, 35–41. [CrossRef]
25. Guillot, A.; Hoyek, N.; Louis, M.; Collet, C. Understanding the timing of motor imagery: Recent findings and future directions. *Int. Rev. Sport Exerc. Psychol.* **2012**, *5*, 3–22. [CrossRef]
26. Zisa, N.S.; Rubio, C.; Gómez, M. Reliability and validity of the mental evocation of images, movements and activities questionnaire: A pilot study. *Rehabilitacion* **2020**, *55*, 258–265. [CrossRef]
27. Malouin, F.; Richards, C.L.; Jackson, P.L.; Lafleur, M.F.; Durand, A.; Doyon, J. The Kinesthetic and Visual Imagery Questionnaire (KVIQ) for assessing motor imagery in persons with physical disabilities: A reliability and construct validity study. *J. Neurol. Phys. Ther.* **2007**, *31*, 20–29. [CrossRef]
28. Kilteni, K.; Andersson, B.J.; Houborg, C.; Ehrsson, H.H. Motor imagery involves predicting the sensory consequences of the imagined movement. *Nat. Commun.* **2018**, *9*, 1–9. [CrossRef]

29. Ruggirello, S.; Campioni, L.; Piermanni, S.; Sebastiani, L.; Santarcangelo, E.L. Does hypnotic assessment predict the functional equivalence between motor imagery and action? *Brain Cogn.* **2019**, *136*, 103598. [CrossRef]
30. Williams, S.E.; Cumming, J.; Edwards, M.G. The functional equivalence between movement imagery, observation, and execution influences imagery ability. *Res. Q. Exerc. Sport* **2011**, *82*, 555–564. [CrossRef]
31. Sirigu, A.; Cohen, L.; Duhamel, J.; Pillon, B.; Dubois, B.; Agid, Y.; Pierrot-Deseilligny, C. Congruent unilateral impairments for real and imagined hand movements. *Neuroreport* **1995**, *6*, 997–1001. [CrossRef]
32. Sirigu, A.; Duhumel, J.-R.; Cohen, L.; Pillon, B.; Dubois, B.; Agid, Y. The Mental Representation of Hand Movements After Parietal Cortex Damage. *Science* **1996**, *273*, 1564–1568. [CrossRef]
33. Olsson, C.J.; Nyberg, L. Motor imagery: If you can't do it, you won't think it. *Scand. J. Med. Sci. Sports* **2010**, *20*, 711–715. [CrossRef]
34. Aglioti, S.M.; Cesari, P.; Romani, M.; Urgesi, C. Action anticipation and motor resonance in elite basketball players. *Nat. Neurosci.* **2008**, *11*, 1109–1116. [CrossRef]
35. Beilock, S.L.; Gonso, S. Putting in the mind versus putting on the green: Expertise, performance time, and the linking of imagery and action. *Q. J. Exp. Psychol.* **2008**, *61*, 920–932. [CrossRef]
36. Ferrer González, B.M.; Echevarría Ruiz de Vargas, C.; Zarco Periñán, M.J.; Docobo Durántez, F. *Adaptacion y Validación al Español de la Escala Fugl-Meyer en el Manejo de la Rehabilitación de Pacientes con ictus*; Universidad de Sevilla: Sevilla, Spain, 2016.
37. Wang, T.N.; Lin, K.C.; Wu, C.Y.; Chung, C.Y.; Pei, Y.C.; Teng, Y.K. Validity, responsiveness, and clinically important difference of the abilhand questionnaire in patients with stroke. *Arch. Phys. Med. Rehabil.* **2011**, *92*, 1086–1091. [CrossRef]
38. Lin, K.C.; Chuang, L.L.; Wu, C.Y.; Hsieh, Y.W.; Chang, W.Y. Responsiveness and validity of three dexterous function measures in stroke rehabilitation. *J. Rehabil. Res. Dev.* **2010**, *47*, 563–572. [CrossRef]
39. Jebsen, R.; Taylor, N.; Trieschmann, R.B.; Trotter, M.; Howard, L. An objective and standardized test of hand function. *Arch. Phys. Med. Rehabil.* **1969**, *50*, 311–319.
40. Bohannon, R.W. Internal Consistency of Dynamometer Measurements in Healthy Subjects and Stroke Patients. *Percept. Mot. Ski.* **1995**, *81*, 1113–1114. [CrossRef]
41. Hall, C.R.; Rodgers, W.M.; Barr, K.A. The Use of Imagery by Athletes in Selected Sports. *Sport Psychol.* **2016**, *4*, 1–10. [CrossRef]
42. Hall, C.; Pongrac, J.; Buckholz, E. The measure of Imagery ability. *Hum. Mov. Sci.* **1985**, *4*, 107–118. [CrossRef]
43. Gregg, M.; Hall, C.; Butler, A. The MIQ-RS: A suitable Option for examining movement imagery ability. *Evid. Based Complement Altern. Med.* **2010**, *7*, 249–257. [CrossRef]
44. Isaac, A.; Marks, D.; Russell, D.G. An instrument for assessing imagery of movement: The Vividness of Movement Imagery Questionnaire (VMIQ). *J. Ment. Imag.* **1986**, *10*, 23–30.
45. Climie, E.A.; Rostad, K. Test Review. Wechsler Adult Intelligence Scale. *J. Psychoeduc. Assess.* **2011**, *29*, 581–586. [CrossRef]
46. Mayes, S.D.; Calhoun, S.L. Wechsler Intelligence Scale for Children-Third and -Fourth Edition predictors of academic achievement in children with attention-deficit/hyperactivity disorder. *Sch. Psychol. Q.* **2007**, *22*, 234–249. [CrossRef]
47. Torres, Y.M.V.; Cárdenas, S.J.; Ramírez, A.R. Adaptation of the behavioral activation for depression scale (BADS) in cancer patients. *Psicooncologia* **2020**, *17*, 25–39.
48. Baztan, J.J.; Perez del Molino, J.; Alarcon, T.; San Cristobal, E.; Izquierdo, G.; Manzarbeitia, J. Indice de Barthel: Instrumento válido para la valoración functional de pacientes con enfermedad cerebrovascular. *Rev. Esp. Geriatr. Gerontol.* **1993**, *28*, 32–40.
49. Vergara, I.; Bilbao, A.; Orive, M.; Garcia-Gutierrez, S.; Navarro, G.; Quintana, J.M. Validation of the Spanish version of the Lawton IADL Scale for its application in elderly people. *Health Q. Life Outcomes* **2012**, *10*, 130. [CrossRef]
50. Crammond, D.J. Motor imagery: Never in your wildest dream. *Trends Neurosci.* **1997**, *20*, 54–57. [CrossRef]
51. Bakker, F.; Boschker, M.; Chung, T. Changes in muscular activity while imagining weight lifting using stimulus or response propositions. *J. Sport Exerc. Psychol.* **1998**, *20*, 313–324. [CrossRef]
52. Sharma, N.; Pomeroy, V.M.; Baron, J.-C. Motor Imagery. A Backdoor to the Motor System After Stroke? *Stroke* **2006**, *37*, 1941–1952. [CrossRef]
53. De Vries, S.; Tepper, M.; Otten, B.; Mulder, T. Recovery of Motor Imagery Ability in Stroke Patients. *Rehabil. Res. Pract.* **2011**, *2011*, 1–9. [CrossRef] [PubMed]
54. Al-Nemr, A.; Abdelazeim, F. Relationship of cognitive functions and gross motor abilities in children with spastic diplegic cerebral palsy. *Appl. Neuropsychol. Child.* **2018**, *7*, 268–276. [CrossRef] [PubMed]
55. Wu, M.; Liang, X.; Lu, S.; Wang, Z. Infant motor and cognitive abilities and subsequent executive function. *Infant Behav. Dev.* **2017**, *49*, 204–213. [CrossRef]
56. Cohen, J.A.; Verghese, J.; Zwerling, J.L. Cognition and gait in older people. *Maturitas* **2016**, *93*, 73–77. [CrossRef] [PubMed]
57. Morales-Osorio, M.A.; Mejía Mejia, J. Imaginería motora graduada en el síndrome de miembro fantasma con dolor. *Rev. Soc. Esp. Dolor* **2012**, *19*, 209–216.
58. Wolf, T.J.; Polatajko, H.; Baum, C.; Rios, J.; Cirone, D.; Doherty, M.; McEwen, S. Combined Cognitive-Strategy and Task-Specific Training Affects Cognition and Upper-Extremity Function in Subacute Stroke: An Exploratory Randomized Controlled Trial. *Am. J. Occup. Ther.* **2016**, *70*, 7002290010p1–7002290010p10. [CrossRef]
59. Belda-Lois, J.M.; Mena-Del Horno, S.; Bermejo-Bosch, I.; Moreno, J.C.; Pons, J.L.; Farina, D.; Iosa, M.; Molinari, M.; Tamburella, F.; Ramos, A.; et al. Rehabilitation of gait after stroke: A review towards a top-down approach. *J. Neuroeng. Rehabil.* **2011**, *8*, 66. [CrossRef]

healthcare

MDPI

Article

A Backward Walking Training Program to Improve Balance and Mobility in Children with Cerebral Palsy

Ji-Young Choi [1], Sung-Min Son [1,*] and Se-Hee Park [2]

[1] Department of Physical Therapy, College of Health Science, Cheongju University, Cheongju 28503, Korea; saviorglory@hanmail.net

[2] Department of Physical Therapy, Graduate School of Health Science, Cheongju University, Cheongju 28503, Korea; sehui7688@naver.com

* Correspondence: ssm04177@gmail.com; Tel.: +82-43-229-8636; Fax: +82-43-229-8969

Abstract: Background: We studied the effects of motor tasks using backward walking training on balance and gait functions of children with cerebral palsy. This was a single-blinded, randomized controlled trial with a crossover design conducted at a single facility. Methods: Among 12 children with cerebral palsy, the forward (FWG) ($n = 6$) and backward walking groups (BWG) ($n = 6$) underwent training three times a week for 4 weeks, 40 min a day. After a 6-week break, the crossover training was conducted. Functional walking variables were measured. Time-Up-and-Go (TUG) test, Figure-8 Walk Test (FW8T), and Pediatric Balance Scale (PBS) were used for measuring balance. Results: Both groups showed significant improvement in walking speed, stride length, and step length. The BWG demonstrated significant improvement in walking speed ($p < 0.05$) compared with the FWG. The TUG test, FW8T, and PBS showed significant improvement. After the 4-week intervention, both groups displayed a remarkable decrease in TUG duration and FW8T. Both groups also exhibited improvement in the PBS; more so in the BWG. Conclusions: Backward walking training with motor dual tasks could be a more effective interventional approach than forward walking training to improve balance and walking functions of children with spastic hemiplegia.

Keywords: backward walking; balance; cerebral palsy

Citation: Choi, J.-Y.; Son, S.-M.; Park, S.-H. A Backward Walking Training Program to Improve Balance and Mobility in Children with Cerebral Palsy. *Healthcare* **2021**, *9*, 1191. https://doi.org/10.3390/healthcare9091191

Academic Editors: Grigorios Nasios, Lambros Messinis, Efthimios Dardiotis and Markos Sgantzos

Received: 10 August 2021
Accepted: 7 September 2021
Published: 9 September 2021

Publisher's Note: MDPI stays neutral with regard to jurisdictional claims in published maps and institutional affiliations.

1. Introduction

Cerebral palsy is a non-progressive lesion that is described as a group of permanent disabilities of motor performance caused by injuries in the brain of a developing fetus or infant [1]. Cerebral palsy is accompanied by disturbances in language, the sensory system, and cognitive behavior, or epilepsy. There could be secondary musculoskeletal problems due to muscle weakness, which is attributed to malalignment, limited range of motion, and asymmetric posture [2]. These deformities are negatively influenced by biomechanical movement and can interfere with balance and gait function [3].

The independent gait in children with cerebral palsy has a great effect on the quality of daily life [4]. The major goal of rehabilitation in children with cerebral palsy is the recovery of independent gait. However, children with cerebral palsy have a restricted capacity for movement that results in gait dysfunction (a short stride, slow walking speed, increased swing phase, and postural instability). Thus, it is important to choose an effective training method to improve the balance and gait of children with cerebral palsy. Walking training may be efficient to improve both muscle tone, postural control, and gait function as well as muscle strength, endurance, and coordination of the lower extremity [5,6]. Particularly, dual-task training, which refers to the capacity to control two tasks simultaneously, can be a very effective way to improve the balance and gait function of children with cerebral palsy [7,8].

Walking while performing activities of daily living (ADL) involves the use of inter-joint coordination and the performance of cognitive processes, such as moving things, manipulation of tools, and talking to others [9]. Thus, dual-task training that combines walking while performing other tasks is required. Stroke patients underwent dual-task training while walking to improve their balance and walking, and dual-task training affects their daily life and balancing ability [10]. Chronic stroke patients who lacked confidence improved in their dual-task performance based on walking training, which had a positive effect on their gait and self-efficacy [11]. Dual-task training enhanced ADL and balance in children with ataxic cerebral palsy. They were trained using ring hanging, cup stacking, and blocks and performed motor dual tasks to enhance weight shifting while catching toys in various directions [12]. Therefore, children with cerebral palsy need an intervention program consisting of walking and dual-task training for performing meaningful movements.

Forward-walking training (FWT) is usually conducted, but backward-walking training (BWT) is a more effective intervention in postural conditioning and walking. BWT improves postural balance and walking ability in children with cerebral palsy [13] and was more significantly associated with the speed and step length of stroke patients' than FWT [14]. Based on previous studies, BWT may lead to better functional improvement than FWT. However, several studies have been focused on the positive effects of BWT in stroke patients, but none were related to children with cerebral palsy. Moreover, no studies have investigated the effects of dual-task training while walking backward.

Therefore, this study aimed to identify the effectiveness of BWT combined with motor tasks in children with spastic hemiplegic cerebral palsy. Our objective included determining the effects of this training on balance and gait function in children with spastic hemiplegic cerebral palsy, which was the novelty of this study.

2. Materials and Methods

2.1. Participants

This study was a randomized controlled trial with a crossover design conducted at a single rehabilitation facility. Twelve children (aged 7–14 years) with a diagnosis of spastic hemiplegic cerebral palsy participated in this study. The inclusion criteria were as follows: (1) children with Level I or II cerebral palsy classified according to the Gross Motor Function Classification System (GMFCS) who could move independently without an assistive device; (2) children with a score of 15 or higher in the Cognitive Assessment of Functional Independent Measure (WeeFIM); and (3) those who did not participate in a BWT or a specific training before 6 months.

The study was approved by the Institutional Review Board of Cheong-ju University, and parents of all children provided signed informed consent to participate in the study. The general characteristics of the participants are described in Table 1.

Table 1. General characteristics of the subjects (N = 12).

Variables	M ± SD
Sex (male/female)	5/7
Affected side (Rt./Lt.)	7/5
GMFCS level (I/II)	8/4
Age (years)	10 ± 2.48
Height (cm)	125 ± 9.99
Weight (kg)	27.33 ± 7.35

M ± SD: mean ± standard deviation. Rt.: Right side, Lt.: Left side. GMFCS: Gross Motor Function Classification System.

2.2. Procedures and Intervention

This study used a two-group randomized trial crossover design, and the children were randomly divided into the FWT and the BWT groups. Both groups received 4 weeks of BWT and 4 weeks of FWT (the training order of one group was FWT and BWT, and in the other group was BWT and FWT), with a washout period of 6 weeks during the post-intervention phase of the 1st period and the pre-intervention phase for the 2nd period. The assessments were measured at four time points, that is, during the pre- and post-intervention phases of the 1st and 2nd periods. Of the 12 participants recruited initially, none dropped out during the intervention period (4 weeks), and all 12 subjects (FWT, n = 6; BWT, n = 6) were included in the final analysis. The experimental design of this study is provided in Figure 1.

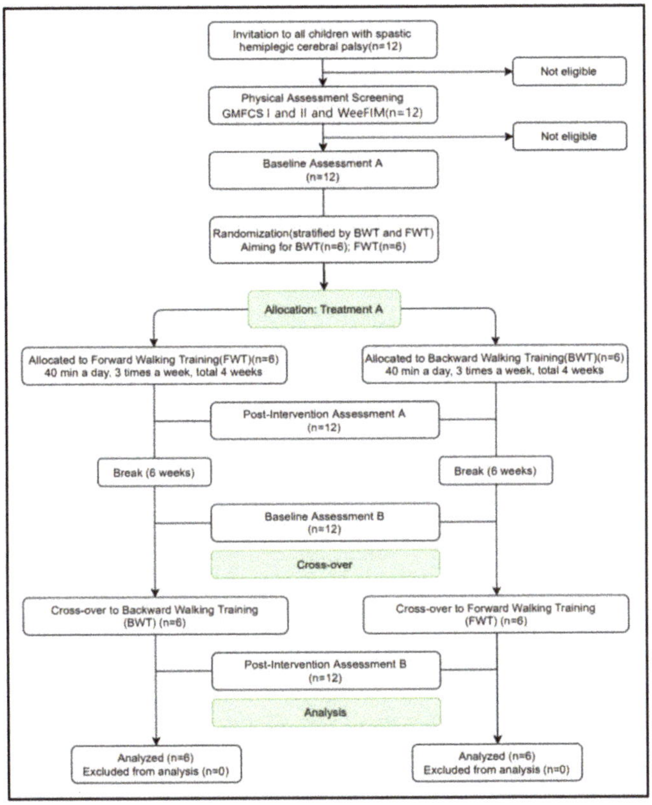

Figure 1. Flowchart in this study. Assessment A, B: Time Up and Go, Figure-8 Walk Test, Pediatric Balance Scale, Opto gait.

2.2.1. Backward Walking Training (BWT)

BWT included gait training on the ground without obstacles. Before BWT, general physical therapy (such as stretching, range of motion exercises, and strengthening exercises) was performed for 10 min. Three sets of BWT with motor dual tasks were executed as per the following schedule: 7 min per set, with approximately 3 min of inter-set rest, for a total training duration of 30 min. The method of BWT applied was based on that described by Davies [15]. First, the therapist helped the participants to move the lower limb correctly when walking backward. The therapist gradually decreased the assistance when the participants underwent training. Second, the participants were then trained to cover a distance of approximately 15 m along the corridor of the treatment room using a safety bar held by the hand on the unaffected side. Third, the participants were encouraged to walk independently without a safety bar. Finally, the participants attempted to walk at a comfortable pace of their choice and gradually increased the distance and speed during backward walking. The training was facilitated by a physical therapist with more than 5 years of experience.

2.2.2. Forward Walking Training (FWT)

During FWT, the therapist assisted in correcting the position of the participant's foot to accurately complete the different elements of the gait cycle. The walking velocity was determined by the pace of the participant. FWT also included the same components, such as general physical therapy and task performance, as BWT.

2.2.3. Motor Task

During the BWT and FWT, the participants were asked to carry a plastic cup containing water (height = 15 cm, base diameter = 6 cm) on a tray. The task was to transport the plastic cup filled with water (using both hands) without spilling its contents, and dropping the cup was considered a failure [14].

2.3. Measurement

2.3.1. Evaluation for Participant Selection

(1) Cross Motor Functional Classification System (GMFCS) GMFCS is a classification system used for measuring the level of motor disorders in children with cerebral palsy. It is divided into five stages according to movements, such as sitting, crawling, and walking, and the degree of mobility using assistive equipment. From level I, which includes children who can walk without any limitations in movement, to level V, which includes difficulty in movements even with assistive equipment; the higher the step, the lower the functional mobility. The inter- and intra-rater reliabilities are 0.93 and 0.97–0.99, respectively [16].
(2) Pediatric Functional Independent Measure for Children (WeeFIM) WeeFIM is a tool used to assess a child's functional ability based on their health, development, education, and social conditions. It is divided into areas of motor and cognitive function, classified into six lower measures, and evaluated using 18 items. Among them, communication, and social cognition are evaluated under the cognitive function. Communication is evaluated for possible comprehension and expression and social interactions. Problem-solving skills and memory are evaluated under social cognition. The evaluation method is conducted by direct observation and interviews by the therapist, and each item is scored on a 7-point scale from 1 to 7. The validity and inter-rater reliability are excellent (intraclass correlation coefficients > 0.90) [17,18].

2.3.2. Outcome Measures

(1) Time-Up-and Go Test TUG is a test that can quickly measure dynamic balance and functional mobility over time. It comprises measuring the time taken from getting up from an armchair, walking 3 m, turning around, and walking back 3 m to sit in place. The participant sits with his/her feet flat on the floor so that the hips and knees are in 90° flexion. The therapist measures the time taken to walk 3 m three times and records the average value. It takes about 11 to 12 s for disabled participants, and if it takes more than 20 s, it is determined that help is needed when walking. For the TUG test, the intra- and interrater reliabilities are ICC = 0.99 and ICC = 0.99, respectively. It is a very reliable measurement method [19].
(2) Figure-8 Walk Test (FW8T) FW8T is a test performed to identify the ability to walk in different paths (straight, curved, clockwise, and counterclockwise) and to recognize the task. The FW8T requires the participant to walk a figure-8 around two cones placed 1.5 m apart. The therapist measures the time taken till the return and the step count. The participant is allowed to practice twice along the path of walking before measurement. The FW8T has excellent test–retest (ICCs = 0.84 and 0.82 for time and steps) and inter-rater reliability (ICCs = 0.90 and 0.92 for time and steps) [20].
(3) Pediatric Balance Scale (PBS) The PBS is developed for school-age children with mild and moderate motor disorders. The PBS assesses the balance and functional ADL and is available for use in children from 5 to 15 years old. It consists of 14 items (with five grade levels) including sitting balance, standing balance, sit-to-stand, stand-to-sit,

moving from chair to chair, standing on one leg, rotating 360 degrees, reaching to the floor, and reaching forward turning. The performance of each task is evaluated on a scale of 0 to 4 points. The PBS score is calculated as static (6 items), dynamic (8 items), and total components (total score), with a maximum total score of 56. The higher the score, the better the balance [21]. The PBS has shown excellent intra-class correlation coefficient (ICC > 0.9) and inter-rater reliability (ICC > 0.9) [22].

(4) Opto Gait We used the Opto Gait system (OPTOGait, Microgate, Bolzano, Italy, 2010) consisting of three transmitting and three receiving bars, to collect data on the participants' walking characteristics. Two bars are placed parallel to each other 1 m apart. Ninety-six LED diodes are positioned on each bar 1 cm apart, 3 mm above the ground. When the participant passes between the transmitting and receiving bars, the system detects the interruption of the optical signal and automatically calculates the spatiotemporal gait parameters based on the presence of a foot in the recording area. The first Opto Gait bar is placed 50 cm from the starting point. The spatiotemporal gait parameters, such as the speed, stride length, and step length of the affected side were analyzed.

2.4. Statistical Analysis

The results of both the interventions, including the 1st and 2nd periods, were pooled. The data were analyzed using SPSS version 22.0 for Windows (IBM, Armonk, NY, USA). To identify the statistically significant differences between pre- and post-test values in the gait and balance function variables (Opto Gait, TUG, FW8T, and PBS) of the BWT and FWT groups, Wilcoxon signed-rank tests were used. The Mann–Whitney U test was used for inter-group comparisons. The statistical significance level was set at 0.05.

3. Results

Twelve participants with spastic CP (5 males, 7 females; mean ± standard deviation age: 10 ± 2.48 years; height = 125 ± 9.99 cm; weight = 27.33 ± 7.35 kg) were included in this study. Participants were either in the GMFCS level I (n = 8) or II (n = 4) (Table 1).

The TUG test scores decreased from 13.85 ± 2.00 and 14.90 ± 1.73 to 13.24 ± 2.24 and 13.33 ± 1.70 in the FWT and the BWT groups, respectively. There were statistically significant differences between the pre-test and post-test assessments in both groups ($p < 0.05$). The BWT group showed a greater reduction than the FWT group, and there was a significant difference in inter-group comparison between the two groups ($p < 0.05$).

In the FW8T, both groups revealed pertinent reductions in the post-test compared to the pre-test evaluations ($p < 0.05$). The difference between the pretest and post-test estimates in the BWT and FWT groups were −1.60 ± 0.48 and −0.70 ± 0.47, respectively. There was a significant difference in inter-group comparison between the two groups ($p < 0.05$).

Both groups displayed a significant increase in the intra-group comparison of PBS ($p < 0.05$). The difference between the pretest and post-test analyses in the BWT and FWT groups were 3.17 ± 0.40 and 1.33 ± 0.51, respectively. There was a significant difference in inter-group comparison between the two groups ($p < 0.05$) (Table 2).

There was a remarkable increase in the intra-group comparison of the spatiotemporal gait parameters (velocity, step, and stride) of both groups ($p < 0.05$). The differences between pre-test and post-test values in the BWT and FWT groups were 0.19 ± 0.11 and 0.06 ± 0.02, respectively. This suggests that only the walking speed increased significantly among the various spatiotemporal gait parameters. However, there were no significant differences in step and stride length ($p > 0.05$) (Table 3).

Table 2. Comparison of changes in balance function in between groups.

		Forward (n = 12)	Backward (n = 12)	U	p (2)
		M ± SD	M ± SD		
TUG	Pre	13.85 ± 2.00	14.90 ± 1.73		
	Post	13.24 ± 2.24	13.33 ± 1.70		
	Post-pre	−0.60 ± 0.59	−1.57 ± 0.31	4.50	0.03
	p (1)	0.02	0.02		
FW8T	Pre	12.20 ± 3.57	12.73 ± 2.74		
	Post	11.50 ± 3.31	11.12 ± 2.45		
	Post-pre	−0.70 ± 0.47	−1.60 ± 0.48	2.00	0.01
	p (1)	0.02	0.02		
PBS	Pre	42.17 ± 5.03	41.33 ± 4.67		
	Post	43.50 ± 4.88	44.50 ± 4.63		
	Post-pre	1.33 ± 0.51	3.17 ± 0.40	0.00	0.00
	p (1)	0.02	0.02		

TUG: Time-Up-and-Go test, FW8T: Figure-8 Walk Test, PBS: Pediatric Balance Scale. (1) Wilcoxon Rank-sum test, (2) Mann–Whitney test.

Table 3. Comparison of changes in gait function in between groups.

		Forward (n = 12)	Backward (n = 12)	U	p (2)
		M ± SD	M ± SD		
Velocity (m/s)	Pre	0.69 ± 0.09	0.66 ± 0.11		
	Post	0.75 ± 0.09	0.85 ± 0.14		
	Post-pre	0.06 ± 0.02	0.19 ± 0.11	2.00	0.00
	p (1)	0.00	0.00		
Step (cm)	Pre	39.46 ± 4.27	38.54 ± 4.10		
	Post	43.29 ± 2.57	43.91 ± 3.62		
	Post-pre	3.83 ± 3.51	5.37 ± 3.55	50.50	0.21
	p (1)	0.00	0.00		
Stride (cm)	Pre	85.38 ± 3.09	87.62 ± 4.55		
	Post	80.50 ± 5.30	80.50 ± 3.42		
	Post-pre	4.87 ± 4.97	7.12 ± 4.83	50.50	0.21
	p (1)	0.00	0.00		

(1) Wilcoxon Rank-sum test, (2) Mann–Whitney test.

4. Discussion

This study aimed to investigate the effects of backward or forward walking combined with task-oriented training on balance ability (TUG, FW8T, and PBS) and functional gait in children with spastic hemiplegic cerebral palsy. The BWT and FWT groups showed significant differences in balance ability and spatiotemporal gait parameters after intervention. In the inter group comparison, the BWT group showed a more significant improvement in balance ability and walking speed parameters than the FWT group. The results of our study showed that backward walking with task-oriented training improved the balancing abilities and functional gait of children with spastic hemiplegic cerebral palsy, more so than those observed in the FWT.

The balancing ability or self-confidence required to improve the gait was higher during BWT because more sensory information is used for balance with limited visual information [23]. In addition, there is higher muscle activity and energy consumption than in FWT [24–26]. Moriello et al. [27] explained that they performed FWT and BWT for patients with spinal cord injury and that the balance and walking confidence improved significantly in the BWT group compared to the FWT group. The improvement in the experimental group is similar to the findings of El-Basatiny et al. [13], who reported that combining the use of BWT and traditional physical therapy in children with hemiparetic cerebral palsy resulted in better improvement in postural balance. Balance and walking combine information from the vestibular, visual, and proprioceptive senses to recognize the body spatially while being involved in posture control [28]. Therefore, backward walking stimulates proprioception, thereby facilitating the activation of righting and equilibrium reactions necessary for maintaining and adjusting posture. All these systems may be associated with the improvement of balance by BWT with dual motor tasks.

In our study, the BWT group showed an enhanced walking speed compared to the FWT group. The increased walking speed in the BWT group may prove the effectiveness of BWT in improving gait capacity. Several mechanisms have been suggested to explain the benefits of BWT in children with spastic cerebral palsy. First, BWT might help increase muscle strength more than FWT. Devita et al. [29] reported that BWT showed a significant increment of muscle activation (quadriceps, hamstring, and tibialis anterior muscle) in stroke patients. An approximate 20% increase in the quadriceps muscle strength was noted during its measurement using a dynamometer. Compared to FWT, BWT resulted in a more eccentric contraction in the quadriceps muscle. Eccentric contraction is more efficient in enhancing muscle strength due to higher energy consumption than concentric contraction [30]. Backward walking induces greater joint movement and ground reaction forces and results in greater energy consumption [31]. Additionally, the interaction of the quadriceps of the knee and flexor digitorum longus of the ankle joint plays an important role in propulsion during the stance phase of backward walking [32]. The momentum gained by the propulsion of the quadriceps during the stance phase might contribute to the enhancement of muscle strength [33]. Second, the BWT influences proprioception, which can contribute to the improvement of gait function. Proprioception is an essential element to maintain balance and serves to control movement when walking [24–26]. Backward walking due to limited vision utilizes proprioception more for the balance response. Although initial motor learning required a high dependence on visual information, proprioceptive information had a greater effect in the learned motor control stage [34]. Mullie et al. [35] reported that stroke patients use their ankle proprioception when walking with their eyes closed. In addition, BWT can induce greater activation of the cerebrum than FWT.

Peters et al. [36] also suggested the backward walking increases the activity of both supplementary and primary motor area of the cerebral cortex compared to forward walking. An analysis of brain activation during backward walking showed that hemodynamic cortical activations in the motor area were greater during backward walking than during forward walking [37]. The improvement in walking speed of children with cerebral palsy could be due to muscle strengthening and activation of proprioceptive information. Therefore, the present study showed a greater positive influence due to BWT with dual motor tasks than due to FWT in children with spastic hemiplegic cerebral palsy.

The limitations of this study are as follows. First, it is difficult to generalize the results for all children with cerebral palsy due to the small number of participants. Second, our study only included children with spastic hemiplegic cerebral palsy with GMFCS levels I and II. Our findings demonstrate that balance function in children with cerebral palsy could be positively influenced by BWT with motor dual task training rather than FWT. Therefore, to redress these limitations, future studies should investigate the results of our study in children with various types of cerebral palsy and use longer washout periods with larger sample sizes.

5. Conclusions

This study aimed to investigate the effect of BWT combined with motor task on balance function and walking capacity of children with CP. Our results demonstrate that children with spastic hemiplegic CP exhibit greater improvements in the balance function (TUG, FW8T, PBS) and walking capacity (gait velocity) following BWT combined with motor tasks than following FWT combined with motor tasks. Our findings suggest effective intervention to promote the ability to balance to prevent fall injuries through BWT and promote self-care activities in daily life. Future studies with larger sample sizes and longer washout periods should explore the effects of BWT and FWT in children with other types of CP as well.

Author Contributions: Conceptualization, J.-Y.C.; Data curation, S.-H.P.; Formal analysis, J.-Y.C. and S.-M.S.; Methodology, S.-M.S.; Writing—original draft, J.-Y.C.; Writing—review and editing, J.-Y.C., S.-M.S. and S.-H.P. Conceptualization, J.-Y.C.; Data curation, J.-Y.C., S.-H.P.; Formal analysis, J.-Y.C., S.-M.S.; Investigation, J.-Y.C.; Methodology, J.-Y.C., S.-M.S.; Supervision, S.-M.S.; Writing—original draft, J.-Y.C.; Writing—review and editing, J.-Y.C., S.-M.S., S.-H.P. All authors have read and agreed to the published version of the manuscript.

Funding: The authors have no financial support for the research.

Institutional Review Board Statement: This study was conducted according to the guidelines of the Declaration of Helsinki and approved by the Institutional Review Board (or Ethics Committee) of Cheong-ju University (Approval code 1041107-202004-HR-010-01, approved on 29 April 2020).

Informed Consent Statement: Informed consent was obtained from all subjects involved in the study.

Data Availability Statement: All of the relevant data are presented within the manuscript. All data are public.

Acknowledgments: The authors would like to thank all the children and their caregivers who cooperated in this experiment to make this study possible.

Conflicts of Interest: The authors declare no conflict of interest.

References

1. Freeman, M. *Physical Therapy of Cerebral Palsy*; Springer Science & Business Media: Berlin, Germany, 2007.
2. Rosenbaum, P.; Paneth, N.; Leviton, A.; Goldstein, M.; Bax, M.; Damiano, D.; Dan, B.; Jacobsson, B. A report: The definition and classification of cerebral palsy April 2006. *Dev. Med. Child Neurol. Suppl.* **2007**, *109*, 8–14.
3. Oeffinger, D.J.; Tylkowski, C.M.; Rayens, M.K.; Davis, R.F.; Gorton, G.E., 3rd; D'Astous, J.; Nicholson, D.E.; Damiano, D.L.; Abel, M.F.; Bagley, A.M.; et al. Gross Motor Function Classification System and outcome tools for assessing ambulatory cerebral palsy: A multicenter study. *Dev. Med. Child Neurol.* **2004**, *46*, 311–319. [CrossRef]
4. Salminen, A.L.; Brandt, A.; Samuelsson, K.; Töytäri, O.; Malmivaara, A. Mobility devices to promote activity and participa-tion: A systematic review. *J. Rehabil. Med.* **2009**, *41*, 697–706. [CrossRef]
5. Roos, P.E.; Barton, N.; van Deursen, R.W. Patellofemoral joint compression forces in backward and forward running. *J. Biomech.* **2012**, *45*, 1656–1660. [CrossRef]
6. Michaelsen, S.M.; Ovando, A.C.; Romaguera, F.; Ada, L. Effect of Backward Walking Treadmill Training on Walking Capacity after Stroke: A Randomized Clinical Trial. *Int. J. Stroke* **2014**, *9*, 529–532. [CrossRef] [PubMed]
7. Blundell, S.W.; Shepherd, R.B.; Dean, C.; Adams, R.D.; Cahill, B.M. Functional strength training in cerebral palsy: A pilot study of a group circuit training class for children aged 4–8 years. *Clin. Rehabil.* **2003**, *17*, 48–57. [CrossRef] [PubMed]
8. Agmon, M.; Belza, B.; Nguyen, H.Q.; Logsdon, R.; Kelly, V.E. A systematic review of interventions conducted in clinical or community settings to improve dual-task postural control in older adults. *Clin. Interv. Aging* **2014**, *9*, 477–492. [CrossRef] [PubMed]
9. Mulder, T.; Zijlstra, W.; Geurts, A. Assessment of motor recovery and decline. *Gait Posture* **2002**, *16*, 198–210. [CrossRef]
10. Haggard, P.; Cockburn, J.; Cock, J.; Fordham, C.; Wade, D. Interference between gait and cognitive tasks in a rehabilitating neurological population. *J. Neurol. Neurosurg. Psychiatry* **2000**, *69*, 479–486. [CrossRef]
11. Page, S.J.; Levine, P.; Sisto, S.; Johnston, M.V. A randomized efficacy and feasibility study of imagery in acute stroke. *Clin. Rehabil.* **2001**, *15*, 233–240. [CrossRef]
12. Bond, J.M.; Morris, M. Goal-directed secondary motor tasks: Their effects on gait in subjects with Parkinson disease. *Arch. Phys. Med. Rehabil.* **2000**, *81*, 110–116. [CrossRef]

13. El-Basatiny, H.M.Y.; Abdel-aziem, A.A. Effect of backward walking training on postural balance in children with hemipa-retic cerebral palsy: A randomized controlled study. *Clin. Rehabil.* **2015**, *29*, 457–467. [CrossRef]
14. Yang, Y.-R.; Wang, R.-Y.; Chen, Y.-C.; Kao, M.-J. Dual-Task Exercise Improves Walking Ability in Chronic Stroke: A Randomized Controlled Trial. *Arch. Phys. Med. Rehabil.* **2007**, *88*, 1236–1240. [CrossRef]
15. Davies, P.M. *Right in the Middle: Selective Trunk Activity in the Treatment of Adult Hemiplegia*; Springer Science & Business Media: Heidelberg, Germany, 1990.
16. Wood, E.; Rosenbaum, P. The gross motor function classification system for cerebral palsy: A study of reliability and stability over time. *Dev. Med. Child Neurol.* **2000**, *42*, 292–296. [CrossRef] [PubMed]
17. Ottenbacher, K.; Taylor, E.T.; Msall, M.E.; Braun, S.; Lane, S.J.; Granger, C.V.; Lyons, N.; Duffy, L.C. The stability and equivalence reliability of the functional independence measure for children (weefim)®. *Dev. Med. Child Neurol.* **2008**, *38*, 907–916. [CrossRef]
18. Sperle, P.A.; Ottenbacher, K.J.; Braun, S.L.; Lane, S.J.; Nochajski, S. Equivalency reliability of the Functional Independence Measure for Children (WeeFIM) administration methods. *Am. J. Occup.* **1997**, *51*, 35–41. [CrossRef] [PubMed]
19. Podsiadlo, D.; Richardson, S. The Timed "Up & Go": A Test of Basic Functional Mobility for Frail Elderly Persons. *J. Am. Geriatr. Soc.* **1991**, *39*, 142–148. [CrossRef]
20. Hess, R.J.; Brach, J.S.; Piva, S.R.; VanSwearingen, J.M. Walking skill can be assessed in older adults: Validity of the Fig-ure-of-8 Walk Test. *Phys. Ther.* **2010**, *90*, 89–99. [CrossRef]
21. Franjoine, M.R.; Gunther, J.S.; Taylor, M.J. Pediatric Balance Scale: A Modified Version of the Berg Balance Scale for the School-Age Child with Mild to Moderate Motor Impairment. *Pediatr. Phys. Ther.* **2003**, *15*, 114–128. [CrossRef]
22. Gan, S.-M.; Tung, L.-C.; Tang, Y.-H.; Wang, C.-H. Psychometric Properties of Functional Balance Assessment in Children with Cerebral Palsy. *Neurorehabilit. Neural. Repair.* **2008**, *22*, 745–753. [CrossRef] [PubMed]
23. Katsavelis, D.; Mukherjee, M.; Decker, L.; Stergiou, N. Variability of lower extremity joint kinematics during backward walking in a virtual environment. *Nonlinear Dyn. Psychol. Life Sci.* **2010**, *14*, 165–178.
24. Winter, D.A.; Pluck, N.; Yang, J.F. Backward Walking: A Simple Reversal of Forward Walking? *J. Mot. Behav.* **1989**, *21*, 291–305. [CrossRef]
25. Foster, H.; DeMark, L.; Spigel, P.M.; Rose, D.K.; Fox, E.J. The effects of backward walking training on balance and mobility in an individual with chronic incomplete spinal cord injury: A case report. *Physiother. Theory Pract.* **2016**, *32*, 536–545. [CrossRef]
26. Rose, D.K.; DeMark, L.; Fox, E.J.; Clark, D.J.; Wludyka, P. A Backward Walking Training Program to Improve Balance and Mobility in Acute Stroke: A Pilot Randomized Controlled Trial. *J. Neurol. Phys. Ther.* **2018**, *42*, 12–21. [CrossRef]
27. Moriello, G.; Pathare, N.; Cirone, C.; Pastore, D.; Shears, D.; Sulehri, S. Comparison of forward versus backward walking using body weight supported treadmill training in an individual with a spinal cord injury: A single subject design. *Physiother. Theory Pract.* **2013**, *30*, 29–37. [CrossRef]
28. Bonan, I.; Marquer, A.; Eskiizmirliler, S.; Yelnik, A.; Vidal, P.-P. Sensory reweighting in controls and stroke patients. *Clin. Neurophysiol.* **2013**, *124*, 713–722. [CrossRef] [PubMed]
29. DeVita, P.; Stribling, J. Lower extremity joint kinetics and energetics during backward running. *Med. Sci. Sports Exerc.* **1991**, *23*, 602–610. [CrossRef] [PubMed]
30. Eilam, D.; Adijes, M.; Vilensky, J. Uphill locomotion in mole rats: A possible advantage of backward locomotion. *Physiol. Behav.* **1995**, *58*, 483–489. [CrossRef]
31. Vilensky, J.; Gankiewicz, E.; Gehlsen, G. A kinematic comparison of backward and forward walking in humans. *J. Hum. Mov. Stud.* **1987**, *13*, 29–50.
32. Grasso, R.; Bianchi, L.; Lacquaniti, F. Motor Patterns for Human Gait: Backward Versus Forward Locomotion. *J. Neurophysiol.* **1998**, *80*, 1868–1885. [CrossRef] [PubMed]
33. Flynn, T.W.; Connery, S.M.; Smutok, M.A.; Zeballos, R.J.; Weisman, I.M. Comparison of cardiopul-monary responses to forward and backward walking and running. *Med. Sci. Sports Exerc.* **1994**, *26*, 89–94. [CrossRef] [PubMed]
34. Shumway-Cook, A.; Woollacott, M. *Motor Control*, 4th ed.; Lippincott Williams & Wilkins: Philadelphia, PA, USA, 2011.
35. Mullie, Y.; Duclos, C. Role of proprioceptive information to control balance during gait in healthy and hemiparetic indi-viduals. *Gait Posture* **2014**, *40*, 610–615. [CrossRef] [PubMed]
36. Peters, S.; Handy, T.C.; Lakhani, B.; Boyd, L.A.; Garland, S.J. Motor and Visuospatial Attention and Motor Planning After Stroke: Considerations for the Rehabilitation of Standing Balance and Gait. *Phys. Ther.* **2015**, *95*, 1423–1432. [CrossRef]
37. Kurz, M.J.; Wilson, T.W.; Arpin, D.J. Stride-time variability and sensorimotor cortical activation during walking. *NeuroImage* **2012**, *59*, 1602–1607. [CrossRef] [PubMed]

healthcare

MDPI

Review

Long-Term Neuropsychological Outcomes Following Temporal Lobe Epilepsy Surgery: An Update of the Literature

Ioanna Alexandratou [1,*], Panayiotis Patrikelis [2], Lambros Messinis [3,4], Athanasia Alexoudi [2], Anastasia Verentzioti [2], Maria Stefanatou [2], Grigorios Nasios [5], Vasileios Panagiotopoulos [6] and Stylianos Gatzonis [2]

1 Department of Neurology, Evangelismos Hospital, Ipsilantou 45-47, 10676 Athens, Greece
2 First Department of Neurosurgery, School of Medicine, National and Kapodistrian University of Athens, 10676 Athens, Greece; ppatrikelis@gmail.com (P.P.); alexoudath@yahoo.gr (A.A.); nverentzioti@gmail.com (A.V.); marwstef@gmail.com (M.S.); sgatzon@med.uoa.gr (S.G.)
3 Lab of Cognitive Neuroscience, Department of Psychology, Aristotle University of Thessaloniki, 54124 Thessaloniki, Greece; lmessinis@psy.auth.gr
4 Department of Psychiatry, University of Patras Medical School, 26504 Patras, Greece
5 Department of Speech and Language Therapy, School of Health Sciences, University of Ioannina, 45500 Ioannina, Greece; grigoriosnasios@gmail.com
6 Department of Neurosurgery, University of Patras Medical School, 26500 Patras, Greece; panagiotopoulos2000@yahoo.com
* Correspondence: io.alexandratou@gmail.com

Abstract: We present an update of the literature concerning long-term neuropsychological outcomes following surgery for refractory temporal lobe epilepsy (TLE). A thorough search was conducted through the PubMed and Medline electronic databases for studies investigating neuropsychological function in adult patients undergoing resective TLE surgery and followed for a mean/median > five years period. Two independent reviewers screened citations for eligibility and assessed relevant studies for the risk of bias. We found eleven studies fulfilling the above requirements. Cognitive function remained stable through long-term follow up despite immediate post-surgery decline; a negative relation between seizure control and memory impairment has emerged and a possible role of more selective surgery procedures is highlighted.

Keywords: refractory temporal seizures; neurosurgery; cognitive outcome; memory; long-term follow-up

Citation: Alexandratou, I.; Patrikelis, P.; Messinis, L.; Alexoudi, A.; Verentzioti, A.; Stefanatou, M.; Nasios, G.; Panagiotopoulos, V.; Gatzonis, S. Long-Term Neuropsychological Outcomes Following Temporal Lobe Epilepsy Surgery: An Update of the Literature. *Healthcare* **2021**, *9*, 1156. https://doi.org/10.3390/healthcare9091156

Academic Editor: Jitendra Singh

Received: 21 July 2021
Accepted: 30 August 2021
Published: 3 September 2021

1. Introduction

Epilepsy surgery is nowadays an evidence-based treatment strategy for patients with drug-resistant epilepsy [1,2]. According to the International League Against Epilepsy (ILAE) definition, pharmacoresistant epilepsy is defined as the failure of a patient's seizures to respond to at least two antiepileptic medications that are appropriately chosen and used for an adequate period [3]. With positive short-term surgery outcomes being definite [4,5], currently, epilepsy surgery centers focus on reporting long-term outcomes from cohort surgical studies implementing a variety of treatment techniques. Reputable reviews have undertaken the task to thoroughly assess post-operative seizure outcome [6,7]. However, it has been long acknowledged that seizure freedom is just one characteristic of surgery outcome. For patients to make truly informed decisions regarding their treatment, they need to know its effect on their ability to work, to study and to socialize. Highlighting the need for more reliable studies reviewing the non-seizure outcomes, neuropsychological data have also been evaluated and much has been documented on long-term postoperative cognitive outcomes [8,9] Although, an association with temporal lobe (TL) surgery and progressive memory decline has been suggested [10,11], while, currently, cognitive function is considered to remain stable one year after surgery [12]. Various factors such as chronological age, seizure recurrence, burden of medication and type of surgery have been linked to long-term postoperative outcomes [13,14]. Despite the encouraging results, epilepsy surgery is still underutilized [15]. Although the importance of early referral has been repeatedly emphasized [16], the delay between a refractory focal epilepsy onset and its surgery still remains of about 15–20 years [17,18]. Admittedly, there are steps to be taken with a view to one of the most frequent chronic and disabling disorders.

We conducted a review of the literature on long-term neuropsychological outcomes following TLE surgery. Our aim was to provide clinicians and researchers with a comprehensive summary reflecting a critical point of view of the current evidence.

2. Methods

2.1. Data Sources

We performed a comprehensive literature search on PubMed with a restriction to full-length English articles published till November 2020, as well as reviews, original articles and book chapters, and consulted experts about other studies. We used the following search terms in various combinations: "refractory temporal seizures", "neurosurgery", "cognitive outcome", "memory" and "long-term follow-up".

2.2. Study Selection and Classification

Two independent reviewers applied the following study inclusion criteria:

1. Reports of >20 patients with a medical history of drug-resistant temporal lobe epilepsy (TLE), undergoing resective surgery;
2. Patients older than 16 years old;
3. A mean/median >5 years post-surgery follow-up;
4. Outcomes explored included long-term postoperative neuropsychological data and possible associated predictive factors (Figure 1).

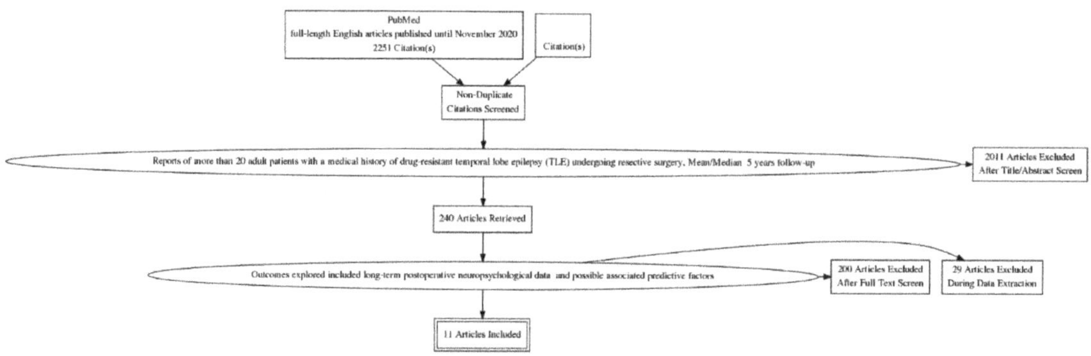

Figure 1. PRISMA flow diagram. The flow diagram depicts the flow of information through the different phases of a systematic review.

3. Results

Eleven studies explored long-term neuropsychological outcomes in adult patients undergoing TL surgery according to the above searching criteria. One study evaluating intelligence consistently reported no worsening of performance following long-term postoperative follow up (Baxendale et al., 2012), while two others reported slight IQ improvement especially in patients achieving seizure freedom (Engman et al., 2006; Alpherts et al., 2004). All studies looked at the long-term memory outcomes, with five showing a greater memory decline following left than right TL resection (Helmstaedter et al., 2003; Rausch et al., 2003; Paglioli et al., 2004; Baxendale et al., 2012; Helmstaedter et al., 2018), and two older ones a progressive cognitive decline and lower memory scores in (Helmstaedter et al., 2003; Rausch et al., 2003). These findings have been challenged by later data, where cognitive function remained stable at one year following surgery, showing no evidence of accelerated memory decline (Engman et al., 2006; Alpherts et al., 2006; Andersson-Roswall et al., 2010; Baxendale et al., 2012; Salvato et al., 2016; Helmstaedter et al., 2018). Four other studies provided an account on which variables influenced postoperative memory improvement (Baxendale et al., 2012; Salvato et al., 2016; Mathon et al., 2017; Helmstaedter et al., 2018), while one study compared selective (anterior temporal lobectomy-ATL) with nonselective (selective amygdalohippocampectomy-SAH) surgery approaches with regard to neuropsychological outcomes, showing that risks of cognitive and/or verbal memory impairment were greater in patients with ATL than in those with SAH (Mathon et al., 2017). (Table 1).

Table 1. Neuropsychological outcome in studies with long-term follow up.

Scheme	Mean Follow-Up Years	Type of Surgery and N Sample Used	Population and Type of Study	Controlled Study	Neuropsychological Outcome
Helmstaedter, 2003	5	Temporal (*N* 147) Medical (*N* 102)	Adults prospective	Yes medical versus surgery	This was greater after a left temporal lobectomy or if seizures continued postoperatively. Seizure-free surgical patients showed a recovery of memory function. Intelligence: No significant changes were seen in either group
Rausch, 2003	12.8	Temporal (*N* 44) Medical (*N* 8)	Adults prospective	Yes medical versus surgery	Memory: Patients with LTL surgery showed selective early decreases in verbal memory. At the long-term follow-up, further decreases in verbal memory and visual memory scores were seen for all patient groups. The nonmemory scores remained stable over time.
Alpherts, 2004	6	Temporal (*N* 71)	Adults prospective	No	Intelligence: Right or left surgery did not affect intelligence
Paglioli, 2004	5.4	Temporal (*N* 65)	Adults prospective	No	Memory: Left side surgery: Of 38 patients, worsening occurred in logical memory in 5 (13%) and in verbal learning in 10 (26%). Right side surgery: Of 27 patients, worsening occurred in logical memory in one (4%), in verbal learning in three (11%), and in visual memory in 6 (22%).
Alpherts, 2006	6	Temporal (*N* 85)	Adults prospective	No	Memory: LTL patients showed an ongoing memory decline for consolidation and acquisition of verbal material for up to 2 years after surgery. RTL patients at first showed a gain in both memory acquisition and consolidation, which vanished in the long term. The group with pure MTS showed an overall lower verbal memory performance than the group without pure MTS (mesiotemporal sclerosis). A dynamic decline of verbal memory functions up to 2 years after left temporal lobectomy, which then levels off.
Engman, 2006	9.8	Temporal (*N* 25) Control group (*N* 25)	Adults prospective	Yes control group versus surgery	Memory: No signs of accelerated cognitive aging after 10 years in a majority of the patients. Those who were seizure-free at long-term follow-up had a significantly higher intelligence score than patients who were still having seizures

Table 1. *Cont.*

Scheme	Mean Follow-Up Years	Type of Surgery and *N* Sample Used	Population and Type of Study	Controlled Study	Neuropsychological Outcome
Andersson-Roswall, 2010	10	Temporal (*N* 51)	Adults prospective	No	Memory: Decline was detected already 2 years postoperatively, with no further decline from 2 to 10 years. The memory decline was not related to seizure outcome or AED treatment.
Baxendale, 2012	9.1	Temporal (*N* 71)	Adults prospective	No	Intelligence: No difference on intellectual function after surgery. Memory: Verbal learning LTL (Left temporal lobe) performed more poorly than the RTL (rlght temporal lobe). Visual learning: Patients who were seizure free at T4 demonstrated a significant improvement in visual learning. Patients who were not seizure free at the long term follow up had experienced a decline in visual learning. Those who were not stable both in verbal and visual memory had more post-operative seizures. Significant role of poor postoperative seizure control in progressive memory impairment suggesting cumulative effect of seizures on memory
Salvato, 2016	5	Temporal (*N* 151)	Adults retrospective	No	Memory: Patients with LTLE worsened in the immediate postsurgical period, their performance progressively improved, and at 5 years after surgery, it returned to be equal to the baseline. Shorter duration of epilepsy, younger age, and withdrawal of AED would predict a better memory outcome
Manton, 2017	8.7	Temporal (*N* 389)	Adults prospective	No	Worsening of cognitive function: Histology ILAE type 2/Preoperative verbal memory deficit/Surgical approach: ATL/Preoperative high seizure frequency/Advanced age at surgery/Surgery on left side/Postoperative major complications/Postoperative depression.
Helmstaedter, 2018		Temporal (*N* 161) Medical (*N* 208)	Adults retrospective	Yes surgery group versus medical group	Memory: In the operated group about 9% demonstrated significant losses in verbal memory, figural memory, or executive functions over the T2-T3 interval. In the nonoperated group 10%, 17%, and 6% showed a decline in verbal memory, figural memory, or executive functions between T1 and T3. 5–22 years after surgery, and compared to baseline, only 17% of those who had undergone left and 10% of those who had right temporal lobe surgery showed losses in verbal memory when they were seizure free, as compared to 37% of patients after left and 20% after right temporal lobe surgery if their seizures continued.

LTL = Left temporal lobe, RTL = Right temporal lobe, MTS = Mesial Temporal sclerosis, AED = antiepileptic drugs, LTLE = Left temporal lobe epilepsy, ATL = Anterior temporal lobectomy.

4. Discussion

It is well established that epilepsy surgery is an excellent treatment option for achieving seizure control in patients suffering refractory TLE. However, risk for memory impairment remains to be considered as a serious post-surgical result. So far, long-term neuropsychological outcomes following TLE surgery have been reported by various prospective studies [19,20]. Given that few reports have focused on the long-term neuropsychological consequences of TLE surgery, we attempted to provide a review of the literature, investigating neuropsychological function of adult patients undergoing resective TLE surgery and followed for a mean/median > five years period.

In 2003, Helmstaedter and associates [10] reviewed cognitive and memory outcomes in 147 surgically- and 102 medically-treated TLE patients. They reported that surgery anticipated decline whether compared to the medically treated group, particularly when performed to the dominant hemisphere (usually the left), or whether seizures continued following surgery. In the same year, Rausch and colleagues [11] evaluated late postoperative cognitive changes in TLE patients undergoing standardized TL resection. Likewise, they reported a progressive cognitive decline continuing 13 years post-surgery, while left (L)TLE patients showed an accelerated memory decline.

In the following years, various longitudinal investigations presented inconsistent findings as to the continuing and accelerated pattern of postoperative memory decline in TLE. Testing for the presence of continuing postoperative verbal memory deficits in TLE patients during a six-year follow-up interval, Alpherts and collaborators [21] firstly provided evidence for a dynamic verbal memory decline up to two years following left temporal lobectomy, which then levels off. Later, similar long-term follow up studies confirmed such findings. Engman and collaborators [22] reported no signs of accelerated cognitive aging for most patients 10 years post-surgery. A longitudinal prospective study further supported the cognitive stability view, whereas the premise of an ongoing progressive verbal memory decline following TL resection was finally declined [23], since no association between seizure outcome and verbal memory course received confirmation.

Since post-surgical cognitive course in general and memory impairment in particular remained an open issue, many authors were willing to identify important determinants, such as postoperative seizure control and age of surgery. By studying the relationship between postoperative memory decline and seizure outcome for over a five-year follow-up period, Baxendale and associates [9] put forward that those who experienced more post-operative seizures presented verbal and visual memory changes, pinpointing to the role of poor seizure control in progressive memory impairment [9]. Similarly, a further risk for postoperative memory decline was poor seizures control [11]. During a five-year post-surgery follow-up, 50–60% of patients suffered some verbal and figural memory loss, with long-term memory gains being less common (15%) after TL surgery. The cumulative effect of seizures on memory was similarly highlighted over the next years by showing that apart from seizure control, shorter epilepsy duration, younger age, and antiepileptic drugs (AEDs) withdrawal would predict a better memory outcome [14]. Others [12] suggested that major losses appear in the early postoperative period, at one-year follow up, while a few patients decline further. Precisely, when seizure free, only 17% of those undergoing left and 10% of those undergoing right TL surgery showed verbal memory losses, as compared to 37% with left and 20% with right TL surgery who continued having seizures. In summary, as to post-surgery seizure outcome, recovery is more frequently observed than continuing decline.

The approach to surgery was another crucial factor studied extensively in the recent years. The rational underling elective surgery approaches is avoiding lesions following resective surgery to eloquent areas of the temporal neocortex, not directly involved in seizure generation. Mathon and collegues [14] compared three surgical approaches: anterior temporal lobectomy (ATL), transcortical selective amygdalohippocampectomy (SAH), and transsylvian SAH. They suggested that transcortical SAH tends to minimize cognitive deterioration after surgery, with the other two techniques having similar effects.

As to the optimal extent of surgical resection in TLE, no specific neurosurgical approach seems to outweigh the others in terms of seizures control [24]. A review suggested that in 76.2% of works there was evidence for a better cognitive outcome following elective surgery (e.g., SAH) as to the selective (S)ATL, while 23.8% of them did not find various neurosurgical procedures to differ [25]. Important as it may seem, more research is required to fully evaluate possible interactions between surgery approach and long-term (>five years) neuropsychological outcome.

Models of Cognitive and Memory Prognosis Following Surgery

In the realm of TLE surgery, the concept of cognitive reserve has been applied in two different models of hippocampal functioning (i.e., functional reserve vs. hippocampal adequacy), in relation to the risk for memory impairment following temporal lobectomy (TLY). The functional reserve model claims that the size of memory loss is related to the spare capacity of the contralateral temporal lobe to support memory functions following resection of the abnormal (ipsilateral) one. IAT (Intracarotid Amobarbital Test) injections contralateral to the side of epileptogenesis typically produce memory impairment, whereas in the non-epileptic hemisphere memory function remains intact following injections to its epileptic counterpart [26–30]. A non-significant relationship has been recorded by some studies between the functional reserve of the contralateral-non-epileptic temporal lobe as assessed by the IAT and memory changes following TLY. There is rising evidence that the functional adequacy of the tissue to be resected determines the nature and extent of post-operative memory loss. The majority of patients with significantly intact memory before surgery were adversely affected following TLY [31–33]. Likely, studies on memory functioning performing IAT injections to the non-epileptic hemisphere showed that patients with a good pre-surgery memory performance were at much greater risk for memory loss than those who performed poor at baseline [34–36]. A weak point of the functional adequacy model; however, is that it does not predict mild material-specific memory deficits following TLY. Although the contralateral temporal lobe alone does not determine the probability of memory loss following TL, its functional contribution should not be ignored, especially if we consider ample clinical evidence documenting the devastating consequences for memory following bilateral hippocampal damage [36]. There is strong evidence of an inverse relation between the risk of postoperative memory impairments and the functional adequacy of the surgical temporal lobe, mostly seen with respect to verbal memory and left MTLE patients, rather than the functional reserve of the contralateral hemisphere [37].

Outcome studies in epilepsy surgery have identified several factors that have repeatedly been shown to be predictive of a poor prognosis, including the initial response to pharmacotherapy, the underlying etiology, and a patient's history of seizure frequency [38]. From a neuropsychological point of view, one may suggest that restricting surgery to lesional and nonfunctional tissue should help to minimize the cognitive losses resulting from surgery. On the other hand, the functional adequacy of the to-be-resected brain tissue appears to be a major determinant of the cognitive outcome after surgery [39]. Stimulated by the ongoing discussion on the cognitive advantages of selective epilepsy surgery over extended standard resections in temporal lobe epilepsy, advances in MRI acquisitions, PET, SPECT, simultaneous EEG and functional MRI, and electrical and magnetic source imaging can be used to infer the localization of epileptic foci and assist in the design of intracranial EEG recording strategies [40]. Naturally, the outcome of epilepsy surgery will depend not only on the pre-surgery brain network but also on how the surgery (i.e., its location and extent) will affect the brain network [41]. Understanding how structural network abnormalities relate to seizure and cognitive outcomes after temporal lobe epilepsy (TLE) surgery can improve prediction of surgical outcomes [42]. The current standard for individualized prediction of surgical outcome primarily relies on clinical variables [43]. However, combining multivariate data and predicting post-surgery seizure freedom and cognitive outcome, is crucial to inform clinical management decisions.

5. Conclusions

While neuropsychological outcome studies of long-term follow-up remain scarce, progress has been made through the recent years, thus enabling clinicians reach into some safe conclusions for neurocognition after epilepsy surgery. Through our review of the literature, cognitive stability appears to be a still valid assumption receiving empirical support. It is also acknowledged that whenever seizures are controlled and medication reduced, recovery is more frequently observed than continuing decline. Elegantly implemented selective surgical procedures seem to limit cognitive side effects following surgery. In conclusion, the decision to proceed to surgery remains a highly individualized procedure requiring patient-tailored clinical and theory-based neuropsychological approaches. A continuing growth of evidence will help both physicians and patients with this important decision-making process. Finally, further data in cognitive reserve studies is warranted to contribute both long-term neuropsychological prognosis and rehabilitation following TLE surgery.

Author Contributions: Conceptualization, I.A. and S.G.; methodology, I.A, V.P. and L.M.; investigation, I.A., A.A. and A.V.; writing-original draft preparation, I.A. and M.S.; writing-review and editing, P.P. and S.G. and G.N.; supervision, S.G. All authors have read and agreed to the published version of the manuscript.

Funding: This research did not receive any specific grant from funding agencies in the public, commercial, or not-for-profit sectors.

Institutional Review Board Statement: Not applicable.

Conflicts of Interest: The authors declare no conflict of interest.

References

1. Wiebe, S.; Blume, W.T.; Girvin, J.P.; Eliasziw, M. Effectiveness and Efficiency of Surgery for Temporal Lobe Epilepsy Study Group. A Randomized, Controlled Trial of Surgery for Temporal-Lobe Epilepsy. *N. Engl. J. Med.* **2001**, *345*, 311–318. [CrossRef]
2. Engel, J. Early Surgical Therapy for Drug-Resistant Temporal Lobe Epilepsy. *JAMA* **2012**, *307*, 922–930. [CrossRef]
3. Kwan, P.; Arzimanoglou, A.; Berg, A.T.; Brodie, M.J.; Allen Hauser, W.; Mathern, G.; Moshé, S.L.; Perucca, E.; Wiebe, S.; French, J. Definition of drug resistant epilepsy: Consensus proposal by the ad hoc Task Force of the ILAE Commission on Therapeutic Strategies. *Epilepsia* **2010**, *51*, 1069–1077. [CrossRef]
4. Gunn, F.R.; Miserocchi, A.; McEvoy, A. Epilepsy Surgery. *Pract. Neurol.* **2019**, *20*, 4–14. [CrossRef]
5. Engel, J. Evolution of Concepts in Epilepsy Surgery. *Epileptic Disord.* **2019**, *21*, 391–409.
6. Téllez-Zenteno, J.F.; Dhar, R.; Wiebe, S. Long-term seizure outcomes following epilepsy surgery: A systematic review and meta-analysis. *Brain* **2005**, *128*, 1188–1198. [CrossRef]
7. Jobst, B.C.; Cascino, G.D. Resective Epilepsy Surgery for Drug-Resistant Focal Epilepsy. *JAMA* **2015**, *313*, 285–293. [CrossRef]
8. Téllez-Zenteno, J.F.; Dhar, R.; Hernandez-Ronquillo, L.; Wiebe, S. Long-term outcomes in epilepsy surgery: Antiepileptic drugs, mortality, cognitive and psychosocial aspects. *Brain* **2006**, *130*, 334–345. [CrossRef]
9. Baxendale, S.; Thompson, P.J.; Duncan, J.S. Neuropsychological function in patients who have had epilepsy surgery: A long-term follow-up. *Epilepsy Behav.* **2012**, *23*, 24–29. [CrossRef]
10. Helmstaedter, C.; Kurthen, M.; Lux, S.; Reuber, M.; Elger, C.E. Chronic epilepsy and cognition: A longitudinal study in temporal lobe epilepsy. *Ann. Neurol.* **2003**, *54*, 425–432. [CrossRef]
11. Rausch, R.; Kraemer, S.; Pietras, C.; Le, M.; Vickrey, B.; Passaro, E. Early and late cognitive changes following temporal lobe surgery for epilepsy. *Neurology* **2003**, *60*, 951–959. [CrossRef]
12. Helmstaedter, C.; Elger, C.; Vogt, V. Cognitive outcomes more than 5 years after temporal lobe epilepsy surgery: Remarkable functional recovery when seizures are controlled. *Seizure* **2018**, *62*, 116–123. [CrossRef]
13. Mathon, B.; Bielle, F.; Samson, S.; Plaisant, O.; Dupont, S.; Bertrand, A.; Miles, R.; Nguyen-Michel, V.-H.; Lambrecq, V.; Calderon-Garcidueñas, A.L.; et al. Predictive factors of long-term outcomes of surgery for mesial temporal lobe epilepsy associated with hippocampal sclerosis. *Epilepsia* **2017**, *58*, 1473–1485. [CrossRef] [PubMed]
14. Salvato, G.; Scarpa, P.; Francione, S.; Mai, R.; Tassi, L.; Scarano, E.; Russo, G.L.; Bottini, G. Declarative long-term memory and the mesial temporal lobe: Insights from a 5-year postsurgery follow-up study on refractory temporal lobe epilepsy. *Epilepsy Behav.* **2016**, *64*, 102–109. [CrossRef] [PubMed]
15. Thijs, R.D.; Surges, R.; O'Brien, T.; Sander, J.W. Epilepsy in adults. *Lancet* **2019**, *393*, 689–701. [CrossRef]
16. Bjellvi, J.; Olsson, I.; Malmgren, K.; Ramsay, K.W. Epilepsy duration and seizure outcome in epilepsy surgery. *Neurology* **2019**, *93*, e159–e166. [CrossRef]

17. De Tisi, J.; Bell, G.S.; Peacock, J.L.; McEvoy, A.W.; Harkness, W.F.; Sander, J.; Duncan, J.S. The long-term outcome of adult epilepsy surgery, patterns of seizure remission, and relapse: A cohort study. *Lancet* **2011**, *378*, 1388–1395. [CrossRef]
18. Vakharia, V.N.; Duncan, J.S.; Witt, J.-A.; Elger, C.E.; Staba, R.; Engel, J. Getting the best outcomes from epilepsy surgery. *Ann. Neurol.* **2018**, *83*, 676–690. [CrossRef]
19. Jutila, L.; Immonen, A.; Mervaala, E.; Partanen, J.; Puranen, M.; Kälviäinen, R.; Alafuzoff, I.; Hurskainen, H.; Vapalahti, M.; Ylinen, A. Long term outcome of temporal lobe epilepsy surgery: Analyses of 140 consecutive patients. *J. Neurol. Neurosurg. Psychiatry* **2002**, *73*, 486–494. [CrossRef] [PubMed]
20. Jutila, L.; Äikiä, M.; Immonen, A.; Mervaala, E.; Alafuzoff, I.; Kälviäinen, R. Long-term memory performance after surgical treatment of unilateral temporal lobe epilepsy (TLE). *Epilepsy Res.* **2014**, *108*, 1228–1237. [CrossRef]
21. Alpherts, W.; Vermeulen, J.; Van Rijen, P.C.; Da Silva, F.H.L.; Van Veelen, C.W. Verbal memory decline after temporal epilepsy surgery?: A 6-year multiple assessments follow-up study. *Neurology* **2006**, *67*, 626–631. [CrossRef] [PubMed]
22. Engman, E.; Andersson-Roswall, L.; Samuelsson, H.; Malmgren, K. Serial cognitive change patterns across time after temporal lobe resection for epilepsy. *Epilepsy Behav.* **2006**, *8*, 765–772. [CrossRef]
23. Andersson-Roswall, L.; Engman, E.; Samuelsson, H.; Malmgren, K. Cognitive outcome 10 years after temporal lobe epilepsy surgery: A prospective controlled study. *Neurology* **2010**, *74*, 1977–1985. [CrossRef]
24. Schramm, J. Temporal lobe epilepsy surgery and the quest for optimal extent of resection: A review. *Epilepsia* **2008**, *49*, 1296–1307. [CrossRef]
25. Tanriverdi, T.; Dudley, R.W.R.; Hasan, A.; Al Jishi, A.; Al Hinai, Q.; Poulin, N.; Colnat-Coulbois, S.; Olivier, A. Memory outcome after temporal lobe epilepsy surgery: Corticoamygdalohippocampectomy versus selective amygdalohippocampectomy. *J. Neurosurg.* **2010**, *113*, 1164–1175. [CrossRef] [PubMed]
26. Loring, D.W.; Murro, A.M.; Meador, K.J.; Lee, G.P.; Gratton, C.A.; Nichols, M.E.; Gallagher, B.B.; King, D.W.; Smith, J.R. Wada memory testing and hippocampal volume measurements in the evaluation for temporal lobectomy. *Neurology* **1993**, *43*, 1789. [CrossRef]
27. Perrine, K.; Gershengorn, J.; Brown, E.R.; Choi, I.S.; Luciano, D.J.; Devinsky, O. Material-specific memory in the intracarotid amobarbital procedure. *Neurology* **1993**, *43*, 706. [CrossRef]
28. The Neural Substrate of Memory Impairment Demonstrated by the Intracarotid Amobarbital Procedure. Available online: https://jamanetwork.com/journals/jamaneurology/article-abstract/590582 (accessed on 18 July 2021).
29. Wyllie, E.; Naugel, R.; Awad, I.; Chelune, G.; Lüders, H.; Dinner, D.; Skibinski, C.; Ahl, J.; Portera-Sánchez, A. Intracarotid Amobarbital Procedure: I. Prediction of Decreased Modality-Specific Memory Scores After Temporal Lobectomy. *Epilepsia* **1991**, *32*, 857–864. [CrossRef] [PubMed]
30. Patrikelis, P.; Lucci, G.; Siatouni, A.; Zalonis, I.; Sakas, D.E.; Gatzonis, S. Simulating memory outcome before right selective amygdalohippocampectomy. *Neuropsychol. Rehabil.* **2013**, *23*, 401–415. [CrossRef]
31. Ivnik, R.J.; Sharbrough, F.W.; Laws, E.R. Anterior Temporal Lobectomy for the Control of Partial Complex Seizures: Information for Counseling Patients. *Mayo Clin. Proc.* **1988**, *63*, 783–793. [CrossRef]
32. Chelune, G.J.; Naugle, R.I.; Luders, H.; Awad, I.A. Prediction of cognitive change as a function of preoperative ability status among temporal lobectomy patients seen at 6-month follow-up. *Neurology* **1991**, *41*, 399. [CrossRef]
33. McSweeny, A.J.; Naugle, R.I.; Chelune, G.J.; Lüders, H. "TScores for Change": An illustration of a regression approach to depicting change in clinical neuropsychology. *Clin. Neuropsychol.* **1993**, *7*, 300–312. Available online: https://www.tandfonline.com/doi/abs/10.1080/13854049308401901 (accessed on 18 July 2021). [CrossRef]
34. Verentzioti, A.; Lucci, G.; Sakas, D.E.; Alexoudi, A.; Gatzonis, S.; Kosmidis, M.; Siatouni, A.; Patrikelis, P. Can Neuropsychological Rehabilitation Determine the Candidacy for Epilepsy Surgery? Implications for Cognitive Reserve Theorizing. *J. Neurol. Neurophysiol.* **2017**, *8*, 1–5. [CrossRef]
35. Kneebone, A.C.; Chelune, G.J.; Dinner, D.S.; Naugle, R.I.; Awad, I.A. Intracarotid Amobarbital Procedure as a Predictor of Material-Specific Memory Change After Anterior Temporal Lobectomy. *Epilepsia* **1995**, *36*, 857–865. [CrossRef]
36. Scoville, W.B.; Milner, B. Loss of Recent Memory after Bilateral Hippocampal Lesions. *J. Neurol. Neurosurg. Psychiatry* **1957**, *20*, 11–21. [CrossRef]
37. Quantitative Magnetic Resonance Imaging in Temporal Lobe Epilepsy: Relationship to Neuropathology and Neuropsycho-logical Function—PubMed. Available online: https://pubmed.ncbi.nlm.nih.gov/1514774/ (accessed on 18 July 2021).
38. Englot, D.J.; Chang, E.F. Rates and predictors of seizure freedom in resective epilepsy surgery: An update. *Neurosurg. Rev.* **2014**, *37*, 389–405. [CrossRef]
39. Helmstaedter, C. Cognitive outcomes of different surgical approaches in temporal lobe epilepsy. *Epileptic Disord.* **2013**, *15*, 221–239. [CrossRef] [PubMed]
40. Duncan, J.S.; Winston, G.P.; Koepp, M.J.; Ourselin, S. Brain imaging in the assessment for epilepsy surgery. *Lancet Neurol.* **2016**, *15*, 420–433. [CrossRef]
41. Taylor, P.N.; Sinha, N.; Wang, Y.; Vos, S.B.; de Tisi, J.; Miserocchi, A.; McEvoy, A.W.; Winston, G.P.; Duncan, J.S. The impact of epilepsy surgery on the structural connectome and its relation to outcome. *NeuroImage Clin.* **2018**, *18*, 202–214. [CrossRef] [PubMed]

42. Sinha, N.; Wang, Y.; da Silva, N.M.; Miserocchi, A.; McEvoy, A.W.; de Tisi, J.; Vos, S.B.; Winston, G.P.; Duncan, J.S.; Taylor, P.N. Structural brain network abnormalities and the probability of seizure recurrence after epilepsy surgery. *Neurology* **2020**, *96*, e758–e771. [CrossRef] [PubMed]
43. Development and Validation of Nomograms to Provide Individualised Predictions of Seizure Outcomes after Epilepsy Sur-gery: A Retrospective Analysis—ScienceDirect. Available online: https://www.sciencedirect.com/science/article/abs/pii/S1474442214703254 (accessed on 27 August 2021).

healthcare

MDPI

Review

Repetitive Transcranial Magnetic Stimulation in the Treatment of Alzheimer's Disease and Other Dementias

Athina-Maria Aloizou [1], Georgia Pateraki [1], Konstantinos Anargyros [1], Vasileios Siokas [1], Christos Bakirtzis [2], Markos Sgantzos [1], Lambros Messinis [3], Grigorios Nasios [4], Eleni Peristeri [1], Dimitrios P. Bogdanos [5], Triantafyllos K. Doskas [6], Georgios Tzeferakos [7] and Efthimios Dardiotis [1,*]

1 Department of Neurology, Laboratory of Neurogenetics, University Hospital of Larissa, Faculty of Medicine, School of Health Sciences, University of Thessaly Biopolis, Mezourlo Hill, 41100 Larissa, Greece; athena_aloi@yahoo.gr (A.-M.A.); georginapat08@gmail.com (G.P.); konanarg@gmail.com (K.A.); vsiokas@med.uth.gr (V.S.); sgantzosmarkos@gmail.com (M.S.); eperiste@uth.gr (E.P.)
2 Multiple Sclerosis Center, B'Department of Neurology, AHEPA University Hospital, Aristotle University of Thessaloniki, 54636 Thessaloniki, Greece; bakirtzischristos@yahoo.gr
3 Neuropsychology Section, Departments of Neurology and Psychiatry, University Hospital of Patras and University of Patras, Medical School, 26504 Patras, Greece; lmessinis@upatras.gr
4 Department of Speech and Language Therapy, School of Health Sciences, University of Ioannina, 45500 Ioannina, Greece; grigoriosnasios@gmail.com
5 Department of Rheumatology and Clinical Immunology, University General Hospital of Larissa, Faculty of Medicine, School of Health Sciences, University of Thessaly, Viopolis, 40500 Larissa, Greece; bogdanos@uth.gr
6 Department of Neurology, Athens Naval Hospital, Deinokratous 70, 11521 Athens, Greece; doskastr@gmail.com
7 Association for Regional Development and Mental Health (EPAPSY), 15124 Marousi, Greece; g.tzeferakos@epapsy.gr
* Correspondence: edar@med.uth.gr; Tel.: +30-241-350-1137 or +30-697-422-4279

Citation: Aloizou, A.-M.; Pateraki, G.; Anargyros, K.; Siokas, V.; Bakirtzis, C.; Sgantzos, M.; Messinis, L.; Nasios, G.; Peristeri, E.; Bogdanos, D.P.; et al. Repetitive Transcranial Magnetic Stimulation in the Treatment of Alzheimer's Disease and Other Dementias. *Healthcare* **2021**, *9*, 949. https://doi.org/10.3390/healthcare9080949

Academic Editor: Phyo Kyaw Myint

Received: 26 May 2021
Accepted: 17 July 2021
Published: 28 July 2021

Abstract: Dementia is a debilitating impairment of cognitive functions that affects millions of people worldwide. There are several diseases belonging to the dementia spectrum, most prominently Alzheimer's disease (AD), vascular dementia (VD), Lewy body dementia (LBD) and frontotemporal dementia (FTD). Repetitive transcranial magnetic stimulation (rTMS) is a safe, non-invasive form of brain stimulation that utilizes a magnetic coil to generate an electrical field and induce numerous changes in the brain. It is considered efficacious for the treatment of various neuropsychiatric disorders. In this paper, we review the available studies involving rTMS in the treatment of these dementia types. The majority of studies have involved AD and shown beneficial effects, either as a standalone, or as an add-on to standard-of-care pharmacological treatment and cognitive training. The dorsolateral prefrontal cortex seems to hold a central position in the applied protocols, but several parameters still need to be defined. In addition, rTMS has shown potential in mild cognitive impairment as well. Regarding the remaining dementias, research is still at preliminary phases, and large, randomized studies are currently lacking.

Keywords: rTMS; dementia; Alzheimer's disease; magnetic stimulation; non-pharmacological treatments

1. Introduction

Dementia is a serious health issue around the globe, with a huge social and economic burden, since it affects a large proportion of an otherwise possibly healthy population that is steadily rendered incapable of self-care [1]. There are several dementia types or/and diseases that can cause dementia, such as Alzheimer's disease (AD), vascular dementia (VD), Lewy body dementia (LBD) and frontotemporal dementia (FTD). Pharmacological treatments for these diseases have been more or less ineffective in halting disease progress and ameliorating symptoms, and as such, other non-pharmacological treatment options are now being explored [2].

Transcranial magnetic stimulation (TMS) was introduced as a non-invasive brain stimulation technique by Barker et al. [3]. In TMS, a magnetic coil over the skull generates a high-intensity pulse which can stimulate neurons, and the stimulation varies according to several technical parameters of the coil and the protocol applied [4]. Single and paired stimuli (single- and paired-pulse TMS) are usually used for physiological/pathophysiological studies, while a series of repetitive stimuli (repetitive TMS, rTMS) can induce alterations that persist in the brain. rTMS has found application in various therapeutic protocols and is being tested in several neurodegenerative diseases, with cognitive sequelae as well, such as multiple sclerosis [5].

rTMS induces changes and influences neuronal blood circulation, metabolism and excitability in the stimulated region and other regions connected to the stimuli focus [6–8]. Its effects can be either excitatory or inhibitory, depending on several rTMS protocol parameters, such as frequency, duration and intensity, as expressed by a percentage of the motor threshold (MT, active or resting, AMT/RMT). The main categorization is based on frequency, with two main rTMS types: low-frequency rTMS (LF-rTMS) ($\leq$1 Hz), known to produce inhibitory results, and high-frequency rTMS (HF-rTMS) (>1 Hz), with excitatory results [9]. The reported aftereffects following the repetitive stimulation are considered to reflect synaptic modulations, based on the principles of long-term potentiation (LTP) and long-term depression (LTD), the balance of which is implicated in important cognitive functions, memory included [8,9].

The aim of this review is to summarize the available literature on the therapeutic application of rTMS in the most frequent dementia types and to discuss how the scientific community should proceed in future studies. Additionally, we present ongoing studies on the matter, whose results are expected in the future. To the best of our knowledge, this is the first review to cover the subject of rTMS in the entirety of the dementia subtypes.

2. Alzheimer's Disease

Alzheimer's disease (AD) is the commonest form of dementia in older populations; in fact, almost 5% of people under 65 are affected, with this percentage rising considerably with age, reaching 40–50% in those aged 85 and over [2]. It is considered a neurodegenerative disease, with cognitive impairment, mainly regarding memory and orientation, and behavioral disorders being the most frequently reported symptoms [2]. A plethora of genetic [10,11] and environmental factors [12] have been implicated in its pathogenesis, with no definitive causative factor having been identified so far. Its pathological hallmarks are extracellular amyloid-beta plaques, also known as senile plaques, and neurofibrillary tangles from hyperphosphorylated tau protein [12,13]. A schematic representation can be seen in Figure 1.

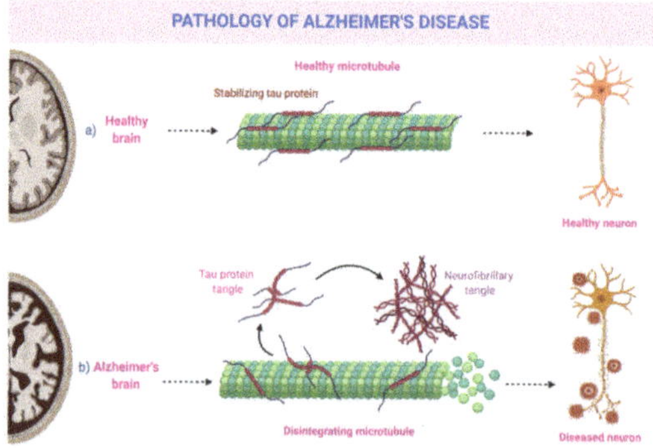

Figure 1. Schematic representation of Alzheimer's disease pathology. The figure was prepared using a BioRender template under license (to DPB).

To date, acetylcholinesterase inhibitors represent the main therapeutic options, but their efficacy in symptom alleviation and disease progress delay is limited [14,15]. As the average life expectancy rises, AD is expected to affect many more millions in the future. It is therefore imperative to find treatments that are actually effective in slowing and possibly reversing its processes, since pharmaceutical agents have proven more or less disappointing, and research has delved into novel pathways involving pathophysiological mechanisms such as endoplasmic reticulum (ER) stress [16]. Non-pharmacological approaches have steadily gained more ground as well, such as cognitive training [17]. In fact, an earlier meta-analysis reported that non-invasive techniques had a significant positive effect on cognitive outcomes [2] and are therefore a promising alternative.

rTMS in the context of dementia has been principally explored in AD, as it is thought to enhance synaptic plasticity, something that can be of utmost importance in preserving cognitive function. As shall be analyzed in detail below, the protocol designs have mostly involved the dorsolateral prefrontal cortex (DLPFC) and regions associated with specific cognitive functions, such as language, memory and attention. Additionally, study designs have been characterized as "online" or "offline" based on whether the stimulation was applied during the course of the cognitive tasks or not, with "online" designs seemingly producing stronger results [2]. Below, we have divided the relevant studies into those focusing on the DLPFC, those that combined rTMS and cognitive training, and those that applied stimulation over different areas. DLPFC studies are presented in Table 1, while studies involving cognitive training can be found in Table 2.

Table 1. rTMS over the DLPFC * in Alzheimer's disease.

Reference	Protocol	Results
[18]	HF ** (20 Hz), bilaterally	Improved action naming
[19]	HF (20 Hz), bilaterally	Improved action naming in milder cognitive decline Improved general naming in moderate to severe cognitive decline
[20]	HF (20 Hz), left	Improved auditory sentence comprehension, persisted for 2 months
[21]	HF (10 Hz), left	Improved neuropsychological test scores and daily functioning
[22]	HF (20 Hz)/LF *3 (1 Hz), bilaterally	Improved cognitive function and mood in the HF group, persisted for 3 months
[23]	HF (20 Hz), bilaterally	Improved cognitive function and behaviorImproved word-image association
[24]	HF (20 Hz), left	Improved cognitive function and behavior
[25]	LF (1 Hz), right	Improved episodic memory (non-verbal recognition)

* Dorsolateral prefrontal cortex. ** High-frequency rTMS. *3 Low-frequency rTMS.

Table 2. rTMS over 6 areas of interest * combined with respective cognitive training in AD.

Reference	Protocol	Results
[26]	HF (10 Hz)	Improved cognitive function, persisted in the maintenance period
[14]	HF (10 Hz)	Improved cognitive function, persisted in the maintenance period
[27]	HF (20 Hz)	Improved cognitive function
[28]	HF (10 Hz)	Improved cognitive function, stronger results for milder cognitive decline
[29]	HF (10 Hz)	Improved cognitive function, persisted in the maintenance period
[30]	HF (10 Hz)	Improved cognitive function, persisted at 6 months for those with better baseline scores
[31]	HF (10 Hz)	Improved cognitive function that persisted, specifically for those with better baseline scores
[32]	HF (10 Hz)	Improved cognitive function, persisted in follow-up, no differences between groups receiving real or sham cognitive training

* Left inferior frontal gyrus, left superior temporal gyrus, left and right dorsolateral prefrontal cortex, left and right parietal somatosensory association cortices.

2.1. *Stimulation of the Dorsolateral Prefrontal Cortex*

The DLPFC has been targeted in a number of studies mainly assessing language in AD. Cotelli et al. (2006) assessed the effects of the left and right DLPFC HF-rTMS (20 Hz, 90% MT intensity) on picture naming during stimulation ("online") [18]. They enrolled 15 anomic AD patients, and performed three blocks of naming tasks, one while they stimulated the left side, one for the right and one for sham. They reported that action/verb naming significantly improved during stimulation of both regions, a finding that was not reported for object/noun naming. The authors also claimed that, compared to other studies that showed improved scores only for left-sided stimulation in normal subjects, this bilateral effect in AD patients probably reflects compensating mechanisms that recruit right-sided networks to support naming and other cognitive functions. The same group, two years later (2008), published another "online" study on DLPFC HF-rTMS (same basic parameters) and naming, examining 24 AD patients with various degrees of cognitive decline [19]. Again, they reported improved action- (but not object-) naming after stimulation of both sides for subjects with mild cognitive decline. On the contrary, naming in both types improved in the individuals with moderate to severe decline. In 2011, the same researchers aimed to further examine this region's effect on language performance and enrolled 10 AD patients, divided into two groups, one receiving real HF-rTMS (20 Hz, 100% MT intensity) stimulation for 4 weeks (5 days/week) over the left DLPFC, and one receiving sham for two weeks and then real stimulation for the subsequent two weeks [20]. They assessed the patients ("offline") at two weeks, four weeks and twelve weeks after initiation, so the last follow-up session was 2 months after the stimulation protocol had ended. At two weeks, there was a significant improvement in auditory sentence comprehension in those receiving real rTMS, which persisted when assessed at follow-up. Other cognitive functions such as memory or other language abilities, naming included, did not show any significant effects, unlike their previous research; they attributed this difference to the different (online vs. offline) designs. The researchers also reported that no additional positive effects were noted from the additional two weeks of rTMS. The wealth of data provided by these three studies was hampered by the fact that all the published data originated from the same study group. One year later, Haffen et al. (2012) described a case of a 75-year-old man with AD, under AD treatment and an antidepressant agent, who received 2 weeks (10 sessions) of HF-rTMS (10 Hz, 100% MT intensity) over his left DLPFC [21]. When he was examined one month after the protocol had ended, he showed improvement in the majority of neuropsychological tests, especially in processing speed and episodic memory, while his environment reported that there was a notable improvement in his everyday activities. No adverse effects or new depressive episodes were noted.

Rutherford et al. (2015) conducted a pilot two-stage study on AD patients with either early or advanced disease [23]. The first stage was a double-blinded, crossover study on nine AD patients, who received 13 sessions of HF-rTMS (20 Hz, 90–100% RMT intensity) over the bilateral DLPFC in the span of 4 weeks. The second stage included blocks of two weeks with 10 real sessions every 3 months, as a follow-up, for six patients that also completed the first phase (min. 10 months, max. 19 months). Montreal Cognitive Assessment (MoCA), Alzheimer's Disease Assessment Scale–Cognitive Subscale (ADAS–Cog) and Revised Memory and Behavior Checklist (RMBC) scores were used for cognitive assessment at baseline, 4 weeks after the last session of the first phase and at the follow-up sessions. Due to several methodological pitfalls, such as scheduling issues for the assessments, most of the results did not reach significance levels. However, an improvement in ADAS-Cog and RMBC was noted after real stimulation. Computerized cognitive exercises were also assessed in a subgroup of patients, where those receiving real training also seemed to perform better, with the scores regarding word–image association reaching the significance threshold. Additionally, patients in the early stages showed a greater overall responsiveness to the treatment, and when analyzed alone, their MoCA scores for the first weeks were significantly better when real was compared to sham.

Wu et al. (2015) assessed the effectiveness of HF-rTMS (20 Hz, 80% MT intensity) over the left DLPFC on cognition and behavioral and psychological symptoms accompanying AD in a double-blinded study [24]. They randomized 54 patients with such symptoms to either active or sham stimulation for five days per week for four weeks, alongside their antipsychotic medication. Ultimately, 26 patients from each treatment branch completed the protocol. The patients were assessed by means of the Behavioral Pathology in Alzheimer's Disease Rating Scale (BEHAVE-AD), ADAS-Cog and the Treatment Emergent

Symptom Scale (TESS) before and after the 4-week protocol. Upon controlling for baseline performances, patients receiving active stimulation had significantly better (i.e., decreased) BEHAVE-AD scores, specifically regarding five of the seven subscore scales, namely activity disturbances, diurnal rhythm, aggressiveness, affective disturbances, anxieties and phobias. They further presented significant improvement in ADAS-Cog scores, compared to sham, in all of the assessed domains, namely language, praxis, memory and attention. Regarding behavioral and psychological symptoms, a higher proportion of patients in the active group showed improvement (73.1% vs. 41.7%). This study holds particular value due to its design, and the fact that it included a larger number of patients than other studies in the same field.

What can be easily deduced from these studies is that HF-rTMS is the method of choice for AD protocols, and that LF-rTMS studies are lacking, since LF-rTMS is not thought to produce beneficial results for these patients. In fact, earlier results suggest that LF-rTMS might even lead to deterioration [33]. To validate this notion, Ahmed et al. (2012) enrolled 45 mild to severe AD patients and divided them equally into three groups [22]. The first group received real HF-rTMS over the DLPFC bilaterally (20 Hz, 100% MT intensity), the second group received LF-rTMS (1 Hz, 100% MT), and the third received sham. The right DLPFC and then the left were stimulated, for one session per day for five consecutive days. The patients were assessed with the Mini Mental State Examination (MMSE), the Instrumental Daily Living Activity (IADL) scale and the Geriatric Depression Scale (GDS) before and after the whole intervention, and then after 1 and 3 months ("offline"). In all the assessment time-points after the intervention, the HF-rTMS group exhibited significantly better scores than the sham and the LF-rTMS groups, despite having no difference at baseline. As improvement persisted at 3 months, the results of this study suggest that rTMS may affect cognition in the long-term. It should be mentioned, however, that only patients with mild to moderate AD responded to treatment, since those with severe dementia did not show improvement in any of the treatment arms.

In this context, the matter of the right DLPFC needs to be discussed. The left DLPFC has been the main focus of the aforementioned studies, which either stimulated both cortices or solely the left. Studies have shown that the recruitment of the right DLPFC occurs in individuals with memory deficits [34], but whether this activation reflects effective compensatory mechanisms [35], or mechanisms with negative impact [36] remains contradictory. In this line of thought, Turrizziani et al. (2012) conducted a study involving 100 healthy young individuals and 8 MCI patients (more studies of this kind shall be analyzed in a section to follow) [37]. They applied verbal and non-verbal recognition tasks (episodic memory) in four sets of experiments, and the stimulation was applied before the recognition, as either LF-rTMS (1 Hz, 90% MT) or HF-rTMS in the form of intermittent theta burst stimulation (three pulses of 50 Hz, 80% AMT) and sham. In the first experiment set, 20 subjects were to receive LF-rTMS and sham over the left DLPFC, and 20 subjects over the right DLPFC, and then participate into a non-verbal recognition memory task. In the second set, the same design was maintained, with 40 different participants, but verbal recognition was tested instead. In the third experiment, ten participants received HF-rTMS over the left DLPFC and then received HF-rTMS over the right, and were tested on non-verbal recognition. Finally, on the fourth experiment, the MCI patients received LF-rTMS to the left and right DLPFC and sham, all in different sessions. They were also tested on non-verbal recognition. The results showed that LF-rTMS over the right and not the left DLPFC significantly improved test accuracy compared to sham in both verbal and non-verbal recognition, while HF-rTMS over the right DLPFC significantly decreased accuracy compared to sham. HF-rTMS over the left DLPFC did not improve test accuracy in the healthy subjects. Regarding the MCI patients, all eight showed non-verbal recognition improvement upon LF-rTMS over the right DLPFC, but not the left, like the first experiment in the healthy controls. This study showed that the activation of the right DLPFC probably negatively impacts memory processes, as showed by the improved test results upon inhibitory stimulation of the right DLPFC. The same research group further explored the effects of LF-rTMS over the right DLPFC in AD patients [25]. They conducted two experiments; in the first, 24 mild AD patients received LF-rTMS (1 Hz, 90% MT) and sham over one hemisphere before a non-verbal recognition test, and two weeks later the other hemisphere was stimulated and tested in a similar manner. In the second experiment, 14 AD patients were randomized to receive either real LF-rTMS over the right DLPFC or sham, for two weeks (5 days per week), and were then assessed at the end of the two weeks and two weeks after that (one month from the protocol initiation). They reported that in the first branch, real stimulation over the right DLPFC led to significant improvement in test accuracy compared to sham, while no difference between real and sham was noted for the left DLPFC. In the second branch, real stimulation was shown to improve performance after the two weeks, which was maintained when assessed at one month follow-up.

Taken together, these results suggest that HF-rTMS over the DLPFC might be beneficial regarding language and other cognitive functions for patients, possibly even in more advanced stages. They also highlight the fact that the brain employs several mechanisms to counterbalance impaired functions, which rTMS seems to affect. One theory to explain this effect is based on dopamine, since studies have shown that HF-rTMS over the DLPFC enhances dopamine production in areas such as the caudate nucleus [38]. Additionally, the effects of rTMS over this region seem to persist through time, albeit this being shown less consistently. However, caution must be exercised overall, as rTMS over the prefrontal cortex has been shown to inhibit other processes, such as memory [39], and the implication of the right DLPFC has yet to be elucidated Thus, its effects need to be assessed regarding all cognitive domains, in order to ensure that it does not aggravate other symptoms, and to define optimal parameters and stimulation targets.

2.2. *Combination of HF-rTMS and Cognitive Training*

Another line of mostly "offline" studies has studied the combined effects of rTMS and cognitive training, and has also included the DLPFC. These studies have mostly exploited the set structure of the NeuroAD™ protocol, as described below.

First Bentwich et al. (2011) enrolled eight mild to moderate AD patients (data for only seven of whom were included due to a withdrawal) and administered HF-rTMS (10 Hz, 90% MT intensity) with concomitant cognitive training for 6 weeks (5 days/week), followed by maintenance sessions (2 days/week) for 3 months [26]. The stimulation was applied over six brain regions, pinpointed with MRI in each patient, corresponding to specific cognitive functions: the left inferior frontal gyrus (known as Broca's area) and the left superior temporal gyrus (known as Wernicke's area) for language, the left and right DLPFC for judgment, executive functions and long-term memory, and the left and right parietal somatosensory association cortices for spatial/topographical orientation and "praxis." The stimulation was combined with computerized cognitive training that included specific tasks on the same functions. They then used a variety of different indices, such as the well-known ADAS-Cog, the Clinical Global Impression of Change (CGIC),the MMSE and the Hamilton Depression Scale (HAL-D), examining the patients after 6 weeks and 4.5 months from the initiation of the stimulation sessions. ADAS-Cog and CGIC scores significantly improved at both assessment times, while most of the other indices also improved but without attaining statistical significance. Two years later, the same group conducted a double-blinded controlled study with 15 AD patients [14]. Seven of those received real HF-rTMS (10 Hz, 90–110% MT intensity depending on the region) over the aforementioned regions and cognitive training for 6 weeks (5/week) followed by two weekly sessions for 3 months, whereas eight received sham stimulation alternatively. After 6 weeks of treatment, the ADAS-Cog scores had significantly improved for the real treatment group when compared to sham, and after 4.5 months, they remained improved, while the scores deteriorated for the sham group. CGIC scores also demonstrated significant improvement. In both studies, patients were under treatment (mostly with cholinesterase inhibitors), something that suggests that rTMS can provide additional benefits to pharmacological treatments. Finally, the group published another study on the joint effects of rTMS–cognitive training [29].

They included 30 mild to moderate AD patients that underwent the same 6-week protocol as described before, with the patients receiving cognitive training while the respective area was being stimulated. Afterwards, tests designed to assess the respective cognitive functions were performed. The researchers reported that ADAS-Cog and MMSE scores significantly improved after the treatment when compared to baseline, while approximately 80% of the patients showed improvement with the stimulation. Five of the patients were also summoned for a second round of treatment approximately 10 months after the first, and the prolonged effect of the first treatment round was assessed. After the second round, the cognitive results were the same or even slightly better than the first, showing that patients did not deteriorate in that 10-month interval after the intervention. Of course, as mentioned before, due to the same group conducting these studies, the results need to be regarded with more caution.

Further enhancing the positive impact of combined rTMS and cognitive training, a poster by Brem et al. (2013) refers to patients with mild AD (number not specified) that received either HF-rTMS (20 Hz, 120% MT intensity) over the same six regions (three per session, randomly selected each time) and concomitant cognitive training for 6 weeks, or sham stimulation and training [27]. Within the first month, patients receiving real treatment showed significant ADAS-Cog improvement compared to sham, and non-significant improvement in MMSE and CGIC. The same group recently published the results of a trial involving 34 AD patients, randomized to receive real or sham HF-rTMS (10 Hz, 80% AMT intensity) over the aforementioned six regions, and real or sham computerized cognitive training [32]. The patients were primarily assessed with the ADAS-Cog scale before, 1 week after and 4–6 weeks after the intervention. Additionally, the Geriatric Depression Scale (GDS) and the Clinical Dementia Rating scale (CDR) were evaluated at baseline, and the Clinical Global Impression of Change (ADCS-CGIC) was administered after the stimulation as well. Overall, directly after the intervention, patients in the real/real group showed greater improvement in ADAS-Cog scores compared to real/sham and sham/sham groups, as neither sham groups had any significant improvement. The real/real group also continued to improve in the follow-up period, and no statistically significant differences in improvement were noted between real/real and real/sham, showing how "removing" the effect of cognitive training did not affect improvement. Upon the combination of sham groups versus the real/real group in the analyses, the improvement of ADAS-CGIC scores also became significant, further highlighting the importance of adding rTMS in order to enhance the training's efficiency. Furthermore, the reported changes could not be attributed to a possible effect of the intervention to depression metrics.

Similarly, Lee et al. (2016) randomized 27 mild to moderate AD patients, 18 to real stimulation and 8 to sham (with one withdrawal) [28]. They kept the same protocols as the previously described studies with the six regions receiving HF-rTMS (10 Hz, 90–110% MT intensity) combined with cognitive training for 6 weeks (5/week), and clinical assessment with the same indices at baseline and at 6 weeks. The ADAS-Cog significantly improved in the real stimulation group, while MMSE and CGIC scores also improved. Additionally, the effects were stronger for the mild AD patients, particularly regarding language and memory. Nguyen et al. (2017) published the results of 10 patients receiving five sessions of HF-rTMS (10 Hz, 100% RMT intensity) over the described areas every week for 5 weeks [30]. Patients were assessed immediately after the protocol ended and after 6 months, by means of MMSE, ADAS-Cog and other scales pertaining to caretaker burden, apathy, locomotor activity and patient dependence. Setting a goal of ameliorating short-term memory, a function tied to the DLPFC, the researchers administered additional stimulation–training sessions over this area every day, either left or right. After the treatment, ADAS-Cog scores were significantly improved, but at 6 months, only the patients with the greatest improvement (>13% improved ADAS-Cog scores) had maintained improved scores. Apathy and dependence scores were found significantly improved throughout all the assessment timepoints. However, these results should be interpreted with caution, as no control group was available, and it is possible that the reported amelioration is the result of a placebo effect or a test–retest learning effect.

Finally, Sabbagh et al. (2020) recently published the results of their phase III randomized, double-blinded, sham-controlled, clinical trial regarding this combination of rTMS and cognitive training [31]. They enrolled a fair number of mild to moderate AD patients and in the final analysis, 59 patients for real (with the same aforementioned parameters) and 59 for sham were included. Patients with better cognitive performance at baseline (ADAS-Cog<30) significantly improved after real stimulation compared to sham and compared to those with scores >30. Additionally, patients in the active group maintained their improvement at a 12-week follow-up assessment, while those having received sham returned to baseline scores. CGIC scores were also significantly different between real and sham, favoring real stimulation, at the 12-week follow-up point. It is interesting to note that, in the active group, only 16% showed a deterioration in CGIC scores, compared to 41.8% for sham, while only 11% with baseline ADAS-Cog<30 from the active group deteriorated, compared to 40% of the same subgroup in the sham branch.

In summary, it appears that this particular combination of HF-rTMS and cognitive training is an effective modality, as all available studies reported beneficial effects and positive results which also seem to persist over time. It is therefore very promising for the future; the areas/regions associated with the impaired language or cognitive functions of the patients could be targeted via rTMS while the patient is simultaneously receiving focused cognitive training. Additionally, it can complement medication, probably producing even better results; several of the studies compared the degree of improvement noted with their intervention to the degree of improvement noted with pharmaceutical agents, rTMS or cognitive training alone, as expressed with the same indices through studies, and showed that it was more effective in the combined condition. Nevertheless, more studies, preferably randomized, controlled and double-blinded, are needed, in order to elucidate the specific effects of rTMS and to accurately identify the most suitable stimulation parameters to secure optimal outcomes.

However, this particular combination protocol must be compared to simpler protocols in order to confirm its superiority, since studies only involving the DLPFC have also produced encouraging results. One study addressed this issue by comparing HF-rTMS (5 Hz, 100% MT intensity) over the left DLPFC to the six areas described above [40]. The researchers randomized 10 participants to DLPFC stimulation and 9 to the six areas of stimulation. They assessed the patients after the stimulation protocol and 4 months after its completion, using ADAS-Cog, MMSE, CGI and other scores pertaining to behavioral and depressive symptoms. In both groups, scores were significantly improved directly after the treatment and at follow-up, while no differences between the protocols were noted in any of the scores. Thus, the authors suggested that the beneficial effect stemming from the more complex protocol is mostly the result of the DLPFC stimulation, an area critical to network integration. However, this study did not involve the cognitive training usually combined with the stimulation of these six areas, and whose therapeutic effect should not be underestimated [17]. On the other hand, acknowledging the complexity of the NeuroAD™ protocol, and the evidence to suggest that only stimulation of the DLPFC suffices to produce beneficial results, Bagattini et al. (2020) wished to further study the combination of cognitive training with rTMS over the left DLPFC only [41]. They conducted a randomized, double-blind, sham-controlled study, allocating 27 patients with either amnesic mild cognitive impairment (MCI) or mild to moderate AD to receive cognitive training directly after real rTMS (20 Hz, 100% RMT intensity), and 23 patients to receive cognitive training after sham rTMS. The RehaCom software was used for the computerized cognitive training sessions, which focused on face–name associative memory. The stimulation was administered 5 days per week for 4 weeks and patients were evaluated at baseline, at 4 weeks and at 12 weeks, by means of MMSE, Geriatric Depressive Scale (GDS) and other tests for specific cognitive functions such as memory, language, attention, spatial reasoning and praxis.The cognitive training significantly ameliorated face–name associative memory, while real stimulation provided significant additional benefits to associative memory, with this improvement being greater for patients with milder disease and higher levels of education. The real group also displayed improved non-trained visuospatial reasoning than the sham group, and this improvement was maintained when assessed at 12 weeks. This study showcases how rTMS can be an important add-on to cognitive training, but since both groups received cognitive training, it fails to provide additional information on whether rTMS is beneficial as a standalone treatment, as the studies analyzed before reported. As such, more studies comparing the available rTMS methods, and their interaction with cognitive training, are warranted.

2.3. Other Areas/Protocols

Koch et al. (2018) enrolled subjects with prodromal AD in order to investigate the effect of stimulation over the precuneus, an area of the parietal lobe thought to be implicated in AD-related memory deficits in early disease stages, due to large neuronal network connectivity impairments [42]. In this double-blinded, randomized, sham-controlled study, seven patients received HF-rTMS (20 Hz, 100% RMT intensity) over the precuneus bilaterally for 10 daily sessions in the span of two weeks, and seven patients received sham stimulation. After a two-week period, the patients were crossed over to the other experiment branch. It is of note that this study also used biomarkers to confirm the diagnosis of prodromal AD, and additionally paired TMS with EEG to uncover the neurophysiological effects of their stimulation. Cognitive assessments were performed with the Alzheimer Disease Cooperative Study Preclinical Alzheimer Cognitive Composite, before and after every two-week protocol. Real stimulation significantly improved episodic memory, with no differences noted for other cognitive functions between real and sham. Neurophysiologically, rTMS enhanced functional connections between the precuneus and medial frontal areas. Increased activity in the precuneus is associated with memory retrieval, while decreased activity is shown during memory encoding. This "encoding/retrieval flip" has been shown to suffer in older individuals with amyloid accumulation [43]. As such, precuneus function enhancement via HF-rTMS expectedly led to memory improvement, as shown in this study.

Avirame et al. (2016) employed deep TMS (dTMS), a method that uses a particular type of coil to reach deep cortical regions, in an attempt to stimulate the prefrontal cortex (PFC) of AD patients [44]. They enrolled 11 patients with moderate to severe AD, who received 20 sessions of dTMS (10 Hz, 100–120% MT intensity) over the PFC bilaterally, assessed by means of Mindstreams (MS) and Addenbrooke Cognitive Examination (ACE) scores before and after the stimulation protocol. An improvement was reported for 60% and 77%of the patients in MS and ACE scores respectively, with this improvement approaching significance. Significance was reached when six patients with more severe disease were separately analyzed. Additionally, improvement in visuospatial abilities was significant, with attention and executive function approaching the threshold as well.

Anderkova et al. (2015) performed an interesting study on how brain atrophy impacts the effectiveness of rTMS [45]. They enrolled 20 patients with mild AD and performed three sessions of HF-rTMS (10 Hz, 90% RMT intensity) over the right inferior frontal gyrus (IFG), the right superior temporal gyrus (STG) and a sham stimulation in a randomized order, and with an interval of at least one day before switching to a different branch. The patients were assessed with the Trail Making Test (TMT), the Stroop Test (ST), the Complex Visual Scene Encoding Task (CVSET) and the MMSE before and after each treatment. Significant improvements on the word part of the ST were noted for both STG and IFG stimulations, with IFG also significantly improving TMT performance; this translates into better attention and psychomotor speed. Regarding atrophy, patients exhibited characteristic patterns of atrophy compared to controls, and a specific pattern of gray matter atrophy correlated with the diminished effectiveness of rTMS on word scores of the ST. This shows how several parameters may affect rTMS effectiveness, and as such studies like this help in stratifying patients more likely to be assisted by the intervention.

Zhao et al. (2017) randomized 30 mild to moderate AD patients (17 for real and 13 for sham stimulation), and applied HF-rTMS (20 Hz, intensity not specified) over three brain areas (parietal P3/P4 and posterior temporal T5/T6, third area not specified) in daily sessions for 6 weeks [46]. The patients were assessed before the protocol, immediately afterwards, and 6 weeks later by means of MoCA, ADAS-Cog, MMSE and World Health Organization University of California-Los Angeles, Auditory Verbal Learning Test (WHO-UCLA AVLT) scores. ADAS-Cog, MMSE and WHO-UCLA AVLT scores were significantly improved at 6 weeks after the intervention, while MoCA scores were significantly improved for the mild subgroup. Additionally, ADAS-Cog scores for moderate patients alone did not significantly improve compared to sham.

3. Mild Cognitive Impairment and Aging

Some studies have included subjects with mild cognitive impairment (MCI); in this condition, individuals do present memory impairment, either subjective or objective, but that is not enough to disturb their daily activities or to fulfill the diagnostic criteria for dementia. However, a fair percentage of MCI patients later end up developing dementia, primarily AD [47]. Consequently, studies on this patient subgroup are also of importance, since impairments are present, and this entity may represent an early stage of dementia as well. Two relevant studies have already been described in the previous section.

First, Cotelli et al. (2012) described the case of an 81-year-old man with amnesic MCI [48]. After two online rTMS sessions to pinpoint the location they would consistently stimulate, they found that only stimulation of the left inferior parietal cortex (IPL) improved accuracy in FNAT (Face–Name Association Test) scores. Subsequently, the patient received HF-rTMS (20 Hz, 100% MT intensity) stimulation over that area for 2 weeks (5/week). A significant improvement in FNAT scores was noted upon completion of the 2 weeks, so the patient exhibited better memory functions, and this change was also evident when the patient was assessed at 24 weeks follow-up. Eliasova et al. (2014) randomized 10 amnesic MCI/AD patients into one group receiving real HF-rTMS (10 Hz, 90% MT intensity) over the right IFG and one group receiving sham treatment [49]. The patients were then assessed with the TMT-A and -B (testing visuospatial processing speed and cognitive flexibility skills, respectively), the ST and the CVSET, before and after the stimulation. Significant effects were noted for both parts of TMT, showing improved attention and psychomotor speed.

DrumondMarra et al. (2015) conducted another randomized, double-blinded, controlled study, by randomizing 34 elderly patients with MCI into either receiving ten sessions of active HF-rTMS (10 Hz, 110% MT intensity) over the left DLPFC (15 patients), or sham (19 patients) [50]. The patients were assessed at baseline, right after the intervention and after one month. Everyday memory improvement, measured by the Rivermead Behavioral Memory Test (RBMT), was noted for the active stimulation, which persisted after one month.

Padala et al. (2018) studied the effects of HF-rTMS (10 Hz, 120% MT) over the left DLPFC in nine MCI patients in order to assess its effectiveness on apathy, an important neurobehavioral aspect of several neurodegenerative conditions [51]. Patients were randomized to either real or sham stimulation (5 days per week for 2 weeks) and were then crossed over to the other branch after an interval of one month. Apathy, executive function and cognition were assessed at baseline, after the interventions, and after the interval. Significant improvement in all these domains was noted after real stimulation, suggesting that rTMS is an attractive option for apathy, a condition inherently difficult to handle pharmacologically [52].

In an attempt to elucidate the mechanisms of rTMS's efficacy and its effects on neural network connectivity, Cui et al. (2019) enrolled 25 MCI patients (21 completed the protocol) in their double-blind, sham-controlled study [53]. They targeted the right DLPFC, as part of the so-called 'default mode network' (DMN), a constellation of functionally connected brain areas that seem to be silenced during attention-requiring tasks, and to represent the brain's intrinsic organization [54]. In this study, the patients were randomized to either receive

HF-rTMS (10 Hz, 90% RMT) or sham for two weeks (5 days/week), and to be assessed directly and two months after its completion via MMSE, ACE-III, GDS and other tests for particular cognitive domains. They reported that real stimulation improved immediate and delayed free recall, and this improvement persisted in the follow-up assessment. This study also included fMRI (functional magnetic resonance imaging) to assess activity within areas of the DMN, and reported that subjects with lower activity levels at baseline presented higher responsiveness to treatment.

Finally, two studies enrolled aging individuals; one healthy and one presenting memory impairment. Normal aging entails a plethora of pathological alterations that also resemble AD, such as ER dysfunction [16], and so the effects of rTMS on aging individuals present a certain interest regarding dementia as well.

Kim et al. (2012) enrolled healthy aging individuals, as, per their rationale, the effect of rTMS had not been investigated in this population [55]. Subjects with concurrent pathologies are more frequently involved in the relevant literature, although normal aging is also associated with faultier selection processes and greater attention deficits upon exposure to task-irrelevant stimuli, alongside other impaired cognitive functions [56]. Thus, the researchers assigned eight individuals into a real stimulation group, receiving HF-rTMS (10 Hz) over the left DLPFC for 5 consecutive days, and eight into a sham group. They used the ST for inhibition control assessment, one day before and one day after the intervention. Those receiving real stimulation showed improvement in task performance, showing that rTMS can prove beneficial even in normal aging. Solé-Padullés et al. (2006) enrolled 40 participants over the age of 50, who complained of memory difficulties and had memory performance within the lower normal range (therefore not fulfilling any dementia criteria) [57]. They then randomized them to one group receiving real HF-rTMS over the right and left DLPFC (10 Hz, 80% MT intensity), and one sham group, and assessed them with FNAT, in an offline design. Only those in the real stimulation group showed significant improvement in associative memory. Further analysis with fMRI demonstrated the recruitment of supplementary regions in the right prefrontal and the posterior cortical areas of both hemispheres, implying that rTMS enhances the activation of these additional areas to facilitate memory functions.

Collectively, these studies show that rTMS can be proven beneficial even for healthy individuals or those with mild disturbances, further highlighting its effectiveness in ameliorating cognitive functions. We believe that longitudinal, sham-controlled studies with patients with MCI/memory complaints that follow an rTMS protocol could help determine whether this early intervention is capable of preventing or delaying full-scale dementia.

4. Frontotemporal Dementia

Frontotemporal dementia (FTD) is a frequent dementia type in individuals below the age of 65, and usually leads to death in less than 10 years. It is characterized by neurodegeneration in the frontal and/or temporal lobes, with a wide array of atrophy patterns, and symptom constellations that include personality alterations, behavioral disorders and language and executive function impairments [58,59]. The main recognized subtypes are the behavioral variant (bvFTD), featuring lack of inhibition, compulsive behavior, personality changes, and Primary Progressive Aphasia (PPA), a syndrome that mostly affects language skills. PPA has three recognized subtypes: the non-fluent/agrammatic variant (nfvPPA), the semantic variant-primary progressive aphasia (svPPA), and the logopenic variant (LPPA) [59,60]. Behavioral and psychological symptoms are usually treated with selective serotonin reuptake inhibitors (SSRIs) and atypical antipsychotics, but no treatment is available for the cognitive deficits [58]. As such, non-pharmacological options are also being explored for this dementia.

Finocchiaro et al. (2006) first reported the use of HF-rTMS (20 Hz, 90% MT intensity) on a 60-year-old right-handed PPA patient with bilateral frontotemporal atrophy, more pronounced on the left hemisphere [61]. They administered two sessions of real HF-rTMS over the left PFC, and one session of sham, assessing the patient with several memory and language tests before and after the sessions. Verb production was significantly enhanced after real stimulation. In a similar vein, Trebbastoni et al. (2013) published another case report on PPA, employing deep HF-rTMS (20 Hz, 100% RMT intensity) and sham over the left DLPFC of a right-handed 50-year-old patient with phonological errors, impaired word recall and sentence repetition, alongside perisylvian atrophy and hypoperfusion, all key features of LPPA [62]. He received two consecutive 5-day real rTMS sessions, and two of sham, and was evaluated before and after with a variety of tasks assessing frontal, language and visuospatial functions. A significant improvement was noted for the language domain after real stimulation. These two case reports suggest that in the setting of PPA, rTMS seems to selectively improve language function, which is the function most heavily impaired in PPA.

Antczak et al. (2018) conducted a pilot study on HF-rTMS for FTD, by enrolling nine patients with bvFTD, one with nfvPPA and one with progressive nfvPPA [58]. The patients received 10 sessions of HF-rTMS (10 Hz, 90% RMT intensity) over the bilateral DLPFC in two weeks, and were cognitively and behaviorally assessed before and after the treatment by means of CGIS, the 21-item HDRS, Geriatric Depression Scale (GDS), Frontal Assessment Battery (FAB) and MoCA. After the intervention, total MoCA score, visuospatial performance and Stroop test subscores (reading time and error number) were improved. Additionally, two out of the three patients with mild depression were shown to return to normal, while a patient with severe depression was afterwards classified as mild.

Here, we deem it useful to mention that a larger, randomized, sham-controlled study on the use of a different form of brain stimulation, the transcranial direct current stimulation, has been recently published [63]. Fifty-five patients and 15 presymptomatic individuals were enrolled, and the left prefrontal cortex was targeted. Improvement in clinical scores and behavioral symptoms was noted after the real stimulation in both groups, alongside an increase in intracortical connectivity. This study exceeds the purposes of the current review, but enhances the notion that non-invasive brain stimulation can be a useful modality for FTD.

This preliminary evidence suggests that rTMS may eventually hold an important position in treating FTD. Understandably, a single study that did not include controls and two case reports are less than enough to reach safe conclusions; this field warrants more research, since this disease affects relatively young individuals who are considerably impaired in their daily functions, with no effective pharmacological treatment available.

5. Vascular Dementia

The second commonest dementia in older ages is vascular dementia (VD), which overlaps with AD in many patients. It stems from progressively acquired ischemic, hypoxic or hemorrhagic brain lesions as a result of cardiovascular and cerebrovascular disorders [64]. VD and AD share several risk factors, such as hypertension and diabetes, but can be clinically differentiated by the fact that in VD, executive dysfunction is usually the first to appear, and cognitive performance seems to fluctuate and worsen abruptly, instead of progressively declining, such as in AD. Mood and personality changes are also more severe in VD. It is of note that cholinergic deficits are noted in VD as well, and this possibly explains why cholinesterase inhibitors are also therapeutically used in this disease [64,65].

Despite it being the second commonest form of dementia and havinga similar pathophysiology to stroke, which has extensive rTMS literature [66], studies on the role of rTMS in VD are few.

Two animal models of VD showed that LF-rTMS (0.5 Hz) and HF-rTMS (5 Hz) significantly improved learning and memory, increased the density of cholinergic neurons and BDNF (brain-derived neurotrophic factor) in hippocampal CA1 area [64,67]. Regarding humans, only two cases have been published so far [65]. Two female patients with VD underwent 40 sessions of a commercially available protocol developed in Mexico, which was otherwise not specified and the stimulation parameters could not be found. The patients were assessed at baseline and two months later. The first patient showed a

7-point improvement in the MMSE and, reportedly, no language difficulty and better social interactions. The second patient showed a 10-point improvement in the MMSE score, with better social interactions and daily activity function.

Two studies involved individuals with known cerebrovascular disease that did not otherwise fulfill dementia criteria. In the earlier study [68], seven such patients with mild executive dysfunction were randomized and then crossed-over, to receive either HF-rTMS (10 Hz, 100% MT intensity) over the left DLPFC or the left motor cortex as a control, undergoing one session of each with a 3-day interval between sessions. They were assessed with a variety of neuropsychological tests, such as the TMT and the Stroop test, focused on psychomotor speed, memory and executive functions. The only significant improvement upon stimulation of the DLPFC was reported for the Stroop test, indicating amelioration in processing speed and attention. However, this study included a small number of patients, and a test–retest effect cannot be excluded either. Sedlackova et al. (2011) enrolled seven subjects with MCI of the vascular type without dementia, and tested HF-rTMS (10 Hz, 100% RMT intensity) and LF-rTMS (1 Hz, 100% RMT intensity) over the left DLPFC, and over the motor cortex as a control, in a crossover design [69]. Numerous short neuropsychological tests, such as the TMT, were then administered. No results in cognitive performance were noted for either intervention over the DLPFC.

As such, there is a great paucity of studies on VD, which we hope will be addressed in the near future by more studies providing knowledge currently lacking.

6. Lewy Body Dementia

Lewy body dementia (LBD) is the second commonest neurodegenerative dementia, and includes dementia with Lewy bodies (DLB) and Parkinson's disease (PD) dementia (PDD) [70]. As evident from its name, the disease is pathologically characterized by Lewy body protein aggregations, and its symptoms, besides cognitive impairment, include Parkinsonism, serious behavioral and psychological disorders, vivid and recurrent hallucinations and severe sensitivity to antipsychotics [71]. No disease-modifying treatment is available for these diseases either, and limitations regarding the treatment of behavioral/psychological symptoms have directed scientific interest towards non-invasive methods [70].

RTMS has been extensively studied in the context of PD, and a recent meta-analysis on the effects of rTMS on the cognitive performance of PD patients reported that HF-rTMS over the DLPFC may indeed be beneficial [72]. Due to the similarities between LBD and PD, and the existing literature and evidence on rTMS's efficacy on psychiatric disorders [73], it has long been hypothesized that rTMS could also be a therapeutic option for LBD [71]. However, only one study has involved rTMS in LBD, focusing on depression. In that study, rTMS was evaluated in six LBD patients with drug-resistant depression. The protocol involved daily sessions of LF-rTMS (1 Hz, 110% MT intensity) over the right DLPFC and HF-rTMS (10 Hz, 100% MT intensity) for the left DLPFC for ten days. Patients were assessed with HAL-D before and after the intervention, which was found to significantly improve depressive symptoms [74].

Finally, besides those analyzed in the aforementioned meta-analysis, another recent study explored rTMS in PDD [75]. The researchers randomized 33 PDD patients to receive either HF-rTMS (20 Hz, 90% RMT) (18 patients) over the hand area of both primary motor cortices for two weeks (5 days/week), or sham (15 patients). They further received monthly boosting sessions for 3 months. The patients were assessed with the MoCA, MMSE, CDR and Memory and Executive Screening (MES) and Instrumental Activity of Daily Living (IADL) scales. The rationale of the researchers was that improvement in the ability to move about the environment more freely would aid in improving cognition, and that the primary motor cortex is itself involved in some cognitive tasks, such as movement imagery, attention and language [76]. Only a small positive effect on MMSE, MoCA and IADL scores alongside an improvement in motor function was noted. Additionally, this improvement in cognition was not detected in the follow-up sessions, and improvements in MoCA and

CDR scales significantly correlated with improvements in the motor assessment. As such, it is possible that the recorded positive effects reflect an influence of the motor cortex on cognitive processes, albeit a small one.

7. Ongoing Trials

Searching the clinicaltrials.gov website (last accessed on 7 June 2021) with the keywords "dementia" and "rTMS", 36 results come up. Of these, one employed transcranial direct stimulation and was thus not further assessed, and from the remaining 35, after removing those with published results and those that were irrelevant, 25 remained and are presented in Table 3.

Table 3. Ongoing studies on the use of repetitive transcranial magnetic stimulation in dementia.

NCT Number	Dementia Type	Details
NCT02621424	MCI */AD **	• Last update: May 2021, active—not recruiting • Randomized, crossover, sham-controlled • Target: DLPFC *5 • Outcome: Cognitive score improvement and CSF *6 BDNF *7 levels
NCT01894620	AD	• Last update: February 2021, completed, preliminary results listed • Randomized, crossover, sham-controlled • Outcome: Cognitive score and sleep improvement
NCT02537496	AD	• Last update: February 2019, completed, no results listed • Randomized, sham-controlled • Target: Left DLPFC • Outcome: Executive function/working memory improvement
NCT04562506	AD	• Last update: September 2020, completed • Randomized, sham-controlled, double-blinded • Target: Bilateral DLPFC • Outcome:Cognitivefunctions
NCT03665831	MCT/AD with comorbid MDD *3	• Last update: October 2019, recruiting • Open-label trial • Target: Left DLPFC • Outcome: Emotional/cognitive symptoms
NCT02908815	AD	• Last update: February 2021, recruiting • Randomized, sham-controlled • Target: DLPFC • Outcome: Cognitive functions
NCT01885806	AD-related apathy	• Last update: June 2013, unknown • Randomized, sham-controlled • Target: Left DLPFC • Outcome: Apathy symptoms
NCT04754152	MCI/AD	• Last update: May 2021, recruiting • Randomized, sham-controlled • Outcome: Cognitive functions
NCT04012346	MCI/AD	• Last update: July 2019, unknown • Randomized, double-blinded, sham-controlled • Outcome: Cognitive functions
NCT04042532	Early onset AD	• Last update: April 2021, enrolling by invitation • Randomized, double-blinded, sham-controlled • Target: Left DLPFC • Outcome: Cognitive functions
NCT04555941	MCI/AD	• Last update: October 2020, recruiting • Randomized, triple-blinded, sham-controlled • Outcome: Cognitive functions

Table 3. *Cont.*

NCT Number	Dementia Type	Details
NCT01481961	Early AD	• Last update: March 2015, completed • Open-label trial • Target: Left DLPFC • Outcome: Cognitive functions
NCT03612622	MCI/Early AD	• Last update: February 2021, completed • Randomized, triple-blinded, sham-controlled • Outcome: Associative memory/cognitive and psychological symptoms
NCT04440891	AD	• Last update: April 2021, recruiting • Randomized, double-blinded, sham-controlled • Outcome: Cognitive functions
NCT03270137	AD	• Last update: September2017, unknown • Randomized, single-blinded • Target: Left DLPFC/six region protocol • Outcome: Cognitive functions
NCT04263194	Mild AD	• Last update: August 2020, recruiting • Randomized, triple-blinded, sham-controlled • Target: DMN *8 • Outcome: Cognitive symptoms
NCT03778151	Mild AD	• Last update: February 2021, completed • Randomized, double-blinded, sham-controlled • Target: DMN • Outcome: Cognitive symptoms
NCT04294888	Prodromal and Preclinical AD	• Last update: March 2020, recruiting • Randomized, cross-over, single-blinded, sham-controlled • Target: DMN • Outcome: Associative memory/functional connectivity
NCT04045990	Amnestic MCI/Logopenic PPA *4	• Last update: November 2020, recruiting • Cross-over, single-blinded, sham-controlled • Target: DMN • Outcome: Language/memory
NCT03406429	Agrammatic Non-Fluent PPA/Logopenic PPA	• Last update: March 2021, recruiting • Open-label, cross-over, sham-controlled • Target: Left DLPFC • Outcome: Language/functional connectivity and cortical thickness
NCT04188067	PPA	• Last update: February 2021, recruiting • Open-label, cross-over, sham-controlled • Target: Left DLPFC • Outcome: Language/functional connectivity
NCT04193267	Logopenic PPA	• Last update: March 2021, recruiting • Open-label trial • Target: Left superior temporal gyrus • Outcome: Language
NCT04431401	PPA	• Last update: June 2020, not yet recruiting • Randomized, triple-blinded, sham-controlled • Outcome: Language/functional connectivity
NCT03153540	Agrammatic Non-Fluent PPA	• Last update: January 2021, recruiting • Randomized, cross-over, quadruple-blinded, sham-controlled • Target: Dominant inferior frontal gyrus • Outcome: Safety, tolerability/language/brain function
NCT03448133	PPA	• Last update: June 2020, withdrawn

* Mild Cognitive Impairment. ** Alzheimer's Disease. *3 Major Depressive Disorder *4 Primary Progressive Aphasia. *5 Dorsolateral Prefrontal Cortex. *6 Cerebrospinal Fluid. *7 Brain-Derived Neurotrophic Factor. *8 Default Mode Network.

8. Discussion

RTMS represents a promising modality in treating a plethora of neurological and psychiatric disorders. Per the newest guidelines, it has received B level recommendation for its use in neurodegenerative disorders, namely PD and multiple sclerosis [73]. This is very encouraging, given that AD and dementias share several common elements with these disorders, such as their pathogenetic mechanisms [77,78]. Additionally, these diseases are known for their cognitive sequelae [79,80], and rTMS has been explored as a treatment modality for cognitive decline in this context as well [5,72]. As such, rTMS's efficacy in this setting raises hopes for its application in dementia as well.

In the available literature, the vast majority of rTMS and dementia studies focus on AD. This is reasonable when one considers that is the most frequent dementia type, but the paucity of studies on other dementias highlights the need for additional research regarding these diseases, as many individuals are also heavily affected by them.

Regarding AD in particular, most studies have shown that HF-rTMS is beneficial, as it improves cognitive performance, measured by a variety of scores, and this effect is not only limited to specific domains, but overall daily functional capacity and quality of life. The DLPFC is the main area of interest in AD-and-rTMS literature, as many studies have either focused on it solely, or have included it in multiple-area protocols. However, the superiority of a more complex protocol over one only involving the DLPFC has yet to be proven. A commercially available system that includes rTMS over six areas of interest combined with respective cognitive training has been explored in several studies, with encouraging results [14,26–32], but only one publication compared rTMS over the DLPFC to rTMS over these six regions, and it did not show any difference [40]. This study did not include cognitive training, so whether the benefits of this system stem from the cognitive training only and not rTMS should be further explored.

The right DLPFC represents another "mystery." It most likely negatively impacts cognition and memory, as its inhibition via LF-rTMS led to episodic memory improvement in both healthy and demented individuals [25,37]. It is likely that, in the protocols with both DLPFCs stimulated, the reported positive effects were the result of the left DLPFC being enhanced, which holds the greatest significance regarding language and memory [81], and this enhancement overcame the counteraction of the right DLPFC. However, the study of Cui et al. (2019) [53] reported that HF-rTMS over the right DLPFC led to improved immediate and delayed free recall, something that contradicts the aforementioned findings. Naturally, methodological differences, namely in the protocols and the tests administered, existed between the studies, and there are not many studies to pool together and draw an accurate conclusion.

A number of studies have assessed the combined efficacy of rTMS and cognitive training [14,26–32], but few attempted to compare the two and examine their interaction. Brem et al. (2020), showed that rTMS was crucial for the effects of cognitive training to become significant [32], supporting the notion that rTMS is the main player in improving cognitive performance and that cognitive training works as an add-on to its effects. On the contrary, Bagattini et al. (2020) showed that rTMS over the left DLPFC served as an add-on to cognitive training instead, improving associative memory and further providing a "generalization" effect, where improvement was noted in domains that had not been cognitively trained, regardless of real or sham rTMS allocation [41]. However, this should be interpreted with caution, since this study did not contain a group that received sham cognitive training and sham rTMS, and a learning effect cannot be excluded. Additionally, no studies directly comparing rTMS as a standalone treatment and pharmacological treatments have been conducted. In the majority of studies, the patients were receiving some sort of standard-of-care treatment for AD, so whether rTMS can be considered a monotherapy or an add-on to other treatments still remains a matter of debate, and studies for the immediate comparison of cognitive training, rTMS, medication and their interaction are required.

Another issue that frequently came up is the greater effectiveness that rTMS seems to have when applied at earlier disease stages. Several studies showed that patients in earlier stages (mild disease) had better responsiveness after treatment [22,23,28,41,46]. This finding is reciprocated by Sabbagh et al. (2020), who showed greater improvement for patients with better baseline scores [31]. Additionally, Nguyen et al. (2017) reported that only patients with the highest baseline scores maintained the improvement induced by rTMS in their follow-up assessment [30]. This phenomenon could also reflect the amount of brain atrophy present, since this worsens as the disease progresses, and Anderkova et al. (2015) showed that gray matter atrophy negatively impacted responsiveness to rTMS [45]. As another metric of disease progression, the employment of additional brain areas for specific tasks, can be detected via fMRI, and represents a compensatory mechanism [25,36]. In this regard, Cui et al. (2019) reported that MCI patients with lower baseline DMN activity benefited more from the intervention [53]. Only one study showed that the subgroup with the more severe disease course benefited more from the intervention [44]. However, this study employed deep TMS, which stimulates more deeply but in a less focused manner. As such, this method may fit better for individuals in more advanced stages, where the brain networks are more diffusely damaged. In any case, the available literature seems to agree that patients gain more out of the procedure when this is applied at earlier stages and when cognitive functions are better preserved. This phenomenon is congruent with the fact that rTMS has also proven beneficial for patients with MCI or generic memory complaints without a diagnosis of dementia, as we analyzed in the section above, and is further corroborated by the plethora of ongoing trials involving MCI individuals. However, the recruitment of patients with severe forms of dementia is more challenging than enrolling those at prodromal and early stages, so there is paucity of studies comparing an adequate number of severe and mild AD patients. In any case, the available results highlight the need for a timely and early intervention, so that the cognitive level may be preserved and even ameliorated. The exact protocols and the intervals for maintenance sessions still need to be determined, but rTMS in AD seems to be a very promising treatment option for the future. Additionally, rTMS in MCI is a very attractive research field, as the true potential of rTMS in delaying dementia progression or even ultimately preventing it can be revealed.

In a similar vein, the long-term efficacy of rTMS is another issue that needs to be discussed. Not all of the studies assessed the patients after the stimulation period ended, and those that did set heterogeneous timepoints, spanning from one or two months [20,21], to 3–4 months [22,26,31] to even 6 months [30]. The most ambitious study [23] involved a more longitudinal follow-up, ranging from 10 to 19 months. However, within this timeframe, many patients were lost to follow-up or ceased their sessions. This represents an issue in assessing the long-term efficacy of rTMS in dementia, since the patients are inherently hard to "maintain," given their old age and burden of disease. Additionally, providing follow-up, boosting sessions also represents a challenge since patients need to be brought to the facility with the machinery, which is not always an easy task for the caregivers and the patients. However, the fact remains that even in these relatively short follow-up periods, some benefits from the intervention were maintained [14,22,26,29,31,32]. Two studies [30,31] reported maintenance, or maintenance at higher levels, of the beneficial effects for those with better baseline scores, further corroborating that rTMS is more efficacious in earlier disease stages. One study [29] re-summoned patients 10 months after their initial protocol. Those that participated in this second round were found to have the same or even better results than the first round. This shows how these patients, albeit few, had not deteriorated within this period, and that after an initial intensive protocol, boosting sessions can be set for a later point in time, possibly assisting in adherence to treatment.

The search for studies on rTMS and the remaining dementias has yielded very few results. For instance, it was very surprising to see that only two cases of rTMS having been applied in VD are available. rTMS is a very safe technique, with very few and minor side-effects, which are mostly self-resolved [4]. Bearing that in mind, the lack of data is rather intriguing. Additionally, in almost all of the aforementioned studies, no side-effect was severe enough to lead to the discontinuation of the protocol. As such, one can only stress how important more studies on these common dementia types and this modality are needed.

It is understandable that large-scale studies with considerable patient cohorts are not easily conducted. A significant limitation would be the lack of the proper equipment by the involved institutions. As such, a multicenter study design is recommended, for larger samples to be gathered and for more accurate conclusions to be drawn. Naturally, randomized, sham-controlled studies must be preferred, and it is imperative that clinical trials in the future are of high quality, in order to provide solid evidence for the efficacy of rTMS. Finally, as has been shown, several published trials in various disorders are of suboptimal quality [82,83], a fact that limits the applicability of their results. Indeed, several of the aforementioned studies had several vital parts of their methodology inadequately reported. Therefore, we propose adherence to the CONSORT statement, as a means ofensuring optimal reporting quality and minimizing bias.

Summing up, in light of the available information, it appears that rTMS holds promise in the amelioration of dementia symptomatology, especially in AD. However, the particulars of the "best" protocol have yet to be defined and high quality clinical trials are urgently needed to provide solid evidence inthis direction. Furthermore, research is still in embryonic stages regarding disorders such as LBD, FTD and VD; this is particularly disappointing considering there is a wealth of literature regarding the application of rTMS in the context of similar disorders such as PD and ischemic stroke [73]. Therefore, we hope that future research endeavors will be turned in this direction, which could help improve the lives of millions of patients suffering from dementia via a safe and effective non-pharmacological intervention.

Author Contributions: Conceptualization, E.D., C.B. and M.S.; methodology, A.-M.A., V.S., K.A. and G.P.; project administration, A.-M.A., G.T. and E.D., validation, C.B., E.P., M.S., T.K.D. and D.P.B.; investigation, A.-M.A., K.A. and G.P.; writing—original draft preparation, A.-M.A., G.P., V.S. and K.A.; writing—review and editing, L.M., G.N., E.P., D.P.B. and G.T.; visualization, A.-M.A. and D.P.B.; supervision, E.D. All authors have read and agreed to the published version of the manuscript.

Funding: This research received no external funding.

Conflicts of Interest: The authors declare no conflict of interest.

References

1. Takizawa, C.; Thompson, P.L.; van Walsem, A.; Faure, C.; Maier, W.C. Epidemiological and Economic Burden of Alzheimer's Disease: A Systematic Literature Review of Data across Europe and the United States of America. *J. Alzheimer's Dis.* **2015**, *43*, 1271–1284. [CrossRef]
2. Hsu, W.-Y.; Ku, Y.; Zanto, T.P.; Gazzaley, A. Effects of Noninvasive Brain Stimulation on Cognitive Function in Healthy Aging and Alzheimer's Disease: A Systematic Review and Meta-Analysis. *Neurobiol. Aging* **2015**, *36*, 2348–2359. [CrossRef]
3. Barker, A.T.; Jalinous, R.; Freeston, I.L. Non-Invasive Magnetic Stimulation of Human Motor Cortex. *Lancet* **1985**, *1*, 1106–1107. [CrossRef]
4. Rossi, S.; Hallett, M.; Rossini, P.M.; Pascual-Leone, A. Safety of TMS Consensus Group Safety, Ethical Considerations, and Application Guidelines for the Use of Transcranial Magnetic Stimulation in Clinical Practice and Research. *Clin.Neurophysiol.* **2009**, *120*, 2008–2039. [CrossRef]
5. Aloizou, A.; Pateraki, G.; Anargyros, K.; Siokas, V.; Bakirtzis, C.; Liampas, I.; Nousia, A.; Nasios, G.; Sgantzos, M.; Peristeri, E.; et al. Transcranial Magnetic Stimulation (TMS) and Repetitive TMS in Multiple Sclerosis. *Rev. Neurosci.* **2021**. [CrossRef]
6. Hallett, M. Transcranial Magnetic Stimulation: A Primer. *Neuron* **2007**, *55*, 187–199. [CrossRef] [PubMed]
7. Hoogendam, J.M.; Ramakers, G.M.J.; Di Lazzaro, V. Physiology of Repetitive Transcranial Magnetic Stimulation of the Human Brain. *Brain Stimul.* **2010**, *3*, 95–118. [CrossRef] [PubMed]
8. Nasios, G.; Messinis, L.; Dardiotis, E.; Papathanasopoulos, P. Repetitive Transcranial Magnetic Stimulation, Cognition, and Multiple Sclerosis: An Overview. *Behav. Neurol.* **2018**, *2018*, 8584653. [CrossRef]
9. Klomjai, W.; Katz, R.; Lackmy-Vallée, A. Basic Principles of Transcranial Magnetic Stimulation (TMS) and Repetitive TMS (RTMS). *Ann. Phys. Rehabil. Med.* **2015**, *58*, 208–213. [CrossRef]
10. Siokas, V.; Aslanidou, P.; Aloizou, A.-M.; Peristeri, E.; Stamati, P.; Liampas, I.; Arseniou, S.; Drakoulis, N.; Aschner, M.; Tsatsakis, A.; et al. Does the CD33 Rs3865444 Polymorphism Confer Susceptibility to Alzheimer's Disease? *J. Mol. Neurosci.* **2020**, *70*, 851–860. [CrossRef] [PubMed]

11. Stamati, P.; Siokas, V.; Aloizou, A.-M.; Karampinis, E.; Arseniou, S.; Rakitskii, V.N.; Tsatsakis, A.; Spandidos, D.A.; Gozes, I.; Mitsias, P.D.; et al. Does SCFD1 Rs10139154 Polymorphism Decrease Alzheimer's Disease Risk? *J. Mol. Neurosci.* **2019**, *69*, 343–350. [CrossRef]
12. Aloizou, A.-M.; Siokas, V.; Vogiatzi, C.; Peristeri, E.; Docea, A.O.; Petrakis, D.; Provatas, A.; Folia, V.; Chalkia, C.; Vinceti, M.; et al. Pesticides, Cognitive Functions and Dementia: A Review. *Toxicol. Lett.* **2020**, *326*, 31–51. [CrossRef] [PubMed]
13. Sami, N.; Rahman, S.; Kumar, V.; Zaidi, S.; Islam, A.; Ali, S.; Ahmad, F.; Hassan, M.I. Protein Aggregation, Misfolding and Consequential Human Neurodegenerative Diseases. *Int. J. Neurosci.* **2017**, *127*, 1047–1057. [CrossRef] [PubMed]
14. Rabey, J.M.; Dobronevsky, E.; Aichenbaum, S.; Gonen, O.; Marton, R.G.; Khaigrekht, M. Repetitive Transcranial Magnetic Stimulation Combined with Cognitive Training Is a Safe and Effective Modality for the Treatment of Alzheimer's Disease: A Randomized, Double-Blind Study. *J. Neural Transm.* **2013**, *120*, 813–819. [CrossRef] [PubMed]
15. Jan, A.T.; Azam, M.; Rahman, S.; Almigeiti, A.M.S.; Choi, D.H.; Lee, E.J.; Haq, Q.M.R.; Choi, I. Perspective Insights into Disease Progression, Diagnostics, and Therapeutic Approaches in Alzheimer's Disease: A Judicious Update. *Front. Aging Neurosci.* **2017**, *9*, 356. [CrossRef]
16. Rahman, S.; Archana, A.; Jan, A.T.; Minakshi, R. Dissecting Endoplasmic Reticulum Unfolded Protein Response (UPRER) in Managing Clandestine Modus Operandi of Alzheimer's Disease. *Front. Aging Neurosci.* **2018**, *10*. [CrossRef] [PubMed]
17. Nousia, A.; Siokas, V.; Aretouli, E.; Messinis, L.; Aloizou, A.-M.; Martzoukou, M.; Karala, M.; Koumpoulis, C.; Nasios, G.; Dardiotis, E. Beneficial Effect of Multidomain Cognitive Training on the Neuropsychological Performance of Patients with Early-Stage Alzheimer's Disease. Available online: https://www.hindawi.com/journals/np/2018/2845176/ (accessed on 19 May 2020).
18. Cotelli, M.; Manenti, R.; Cappa, S.F.; Geroldi, C.; Zanetti, O.; Rossini, P.M.; Miniussi, C. Effect of Transcranial Magnetic Stimulation on Action Naming in Patients with Alzheimer Disease. *Arch. Neurol.* **2006**, *63*, 1602–1604. [CrossRef] [PubMed]
19. Cotelli, M.; Manenti, R.; Cappa, S.F.; Zanetti, O.; Miniussi, C. Transcranial Magnetic Stimulation Improves Naming in Alzheimer Disease Patients at Different Stages of Cognitive Decline. *Eur. J. Neurol.* **2008**, *15*, 1286–1292. [CrossRef] [PubMed]
20. Cotelli, M.; Calabria, M.; Manenti, R.; Rosini, S.; Zanetti, O.; Cappa, S.F.; Miniussi, C. Improved Language Performance in Alzheimer Disease Following Brain Stimulation. *J. Neurol. Neurosurg. Psychiatry* **2011**, *82*, 794–797. [CrossRef]
21. Haffen, E.; Chopard, G.; Pretalli, J.-B.; Magnin, E.; Nicolier, M.; Monnin, J.; Galmiche, J.; Rumbach, L.; Pazart, L.; Sechter, D.; et al. A Case Report of Daily Left Prefrontal Repetitive Transcranial Magnetic Stimulation (RTMS) as an Adjunctive Treatment for Alzheimer Disease. *Brain Stimul.* **2012**, *5*, 264–266. [CrossRef]
22. Ahmed, M.A.; Darwish, E.S.; Khedr, E.M.; El Serogy, Y.M.; Ali, A.M. Effects of Low versus High Frequencies of Repetitive Transcranial Magnetic Stimulation on Cognitive Function and Cortical Excitability in Alzheimer's Dementia. *J. Neurol.* **2012**, *259*, 83–92. [CrossRef]
23. Rutherford, G.; Lithgow, B.; Moussavi, Z. Short and Long-Term Effects of RTMS Treatment on Alzheimer's Disease at Different Stages: A Pilot Study. *J. Exp. Neurosci.* **2015**, *9*, 43–51. [CrossRef] [PubMed]
24. Wu, Y.; Xu, W.; Liu, X.; Xu, Q.; Tang, L.; Wu, S. Adjunctive Treatment with High Frequency Repetitive Transcranial Magnetic Stimulation for the Behavioral and Psychological Symptoms of Patients with Alzheimer's Disease: A Randomized, Double-Blind, Sham-Controlled Study. *Shanghai Arch. Psychiatry* **2015**, *27*, 280–288. [CrossRef] [PubMed]
25. Turriziani, P.; Smirni, D.; Mangano, G.R.; Zappalà, G.; Giustiniani, A.; Cipolotti, L.; Oliveri, M. Low-Frequency Repetitive Transcranial Magnetic Stimulation of the Right Dorsolateral Prefrontal Cortex Enhances Recognition Memory in Alzheimer's Disease. *J. Alzheimer's Dis.* **2019**, *72*, 613–622. [CrossRef]
26. Bentwich, J.; Dobronevsky, E.; Aichenbaum, S.; Shorer, R.; Peretz, R.; Khaigrekht, M.; Marton, R.G.; Rabey, J.M. Beneficial Effect of Repetitive Transcranial Magnetic Stimulation Combined with Cognitive Training for the Treatment of Alzheimer's Disease: A Proof of Concept Study. *J. Neural Transm.* **2011**, *118*, 463–471. [CrossRef]
27. Brem, A.-K.; Schilberg, L.; Freitas, C.; Atkinson, N.; Seligson, E.; Pascual-Leone, A. Effects of Cognitive Training and RTMS in Alzheimer's Disease. *Alzheimer's Dement.* **2013**, *9*, P664. [CrossRef]
28. Lee, J.; Choi, B.H.; Oh, E.; Sohn, E.H.; Lee, A.Y. Treatment of Alzheimer's Disease with Repetitive Transcranial Magnetic Stimulation Combined with Cognitive Training: A Prospective, Randomized, Double-Blind, Placebo-Controlled Study. *J. Clin. Neurol.* **2016**, *12*, 57–64. [CrossRef]
29. Rabey, J.M.; Dobronevsky, E. Repetitive Transcranial Magnetic Stimulation (RTMS) Combined with Cognitive Training Is a Safe and Effective Modality for the Treatment of Alzheimer's Disease: Clinical Experience. *J. Neural Transm.* **2016**, *123*, 1449–1455. [CrossRef]
30. Nguyen, J.-P.; Suarez, A.; Kemoun, G.; Meignier, M.; Le Saout, E.; Damier, P.; Nizard, J.; Lefaucheur, J.-P. Repetitive Transcranial Magnetic Stimulation Combined with Cognitive Training for the Treatment of Alzheimer's Disease. *Neurophysiol. Clin.* **2017**, *47*, 47–53. [CrossRef] [PubMed]
31. Sabbagh, M.; Sadowsky, C.; Tousi, B.; Agronin, M.E.; Alva, G.; Armon, C.; Bernick, C.; Keegan, A.P.; Karantzoulis, S.; Baror, E.; et al. Effects of a Combined Transcranial Magnetic Stimulation (TMS) and Cognitive Training Intervention in Patients with Alzheimer's Disease. *Alzheimers Dement.* **2020**, *16*, 641–650. [CrossRef] [PubMed]
32. Brem, A.-K.; Di Iorio, R.; Fried, P.J.; Oliveira-Maia, A.J.; Marra, C.; Profice, P.; Quaranta, D.; Schilberg, L.; Atkinson, N.J.; Seligson, E.E.; et al. Corticomotor Plasticity Predicts Clinical Efficacy of Combined Neuromodulation and Cognitive Training in Alzheimer's Disease. *Front. Aging Neurosci.* **2020**, *12*, 200. [CrossRef] [PubMed]

33. Trojano, L.; Conson, M.; Maffei, R.; Grossi, D. Categorical and Coordinate Spatial Processing in the Imagery Domain Investigated by RTMS. *Neuropsychologia* **2006**, *44*, 1569–1574. [CrossRef] [PubMed]
34. Bai, F.; Zhang, Z.; Watson, D.R.; Yu, H.; Shi, Y.; Yuan, Y.; Zang, Y.; Zhu, C.; Qian, Y. Abnormal Functional Connectivity of Hippocampus during Episodic Memory Retrieval Processing Network in Amnestic Mild Cognitive Impairment. *Biol. Psychiatry* **2009**, *65*, 951–958. [CrossRef] [PubMed]
35. Smith, G.E.; Pankratz, V.S.; Negash, S.; Machulda, M.M.; Petersen, R.C.; Boeve, B.F.; Knopman, D.S.; Lucas, J.A.; Ferman, T.J.; Graff-Radford, N.; et al. A Plateau in Pre-Alzheimer Memory Decline: Evidence for Compensatory Mechanisms? *Neurology* **2007**, *69*, 133–139. [CrossRef] [PubMed]
36. Grady, C.L.; McIntosh, A.R.; Beig, S.; Craik, F.I. An Examination of the Effects of Stimulus Type, Encoding Task, and Functional Connectivity on the Role of Right Prefrontal Cortex in Recognition Memory. *Neuroimage* **2001**, *14*, 556–571. [CrossRef]
37. Turriziani, P.; Smirni, D.; Zappalà, G.; Mangano, G.R.; Oliveri, M.; Cipolotti, L. Enhancing Memory Performance with RTMS in Healthy Subjects and Individuals with Mild Cognitive Impairment: The Role of the Right Dorsolateral Prefrontal Cortex. *Front. Hum. Neurosci.* **2012**, *6*, 62. [CrossRef]
38. Strafella, A.P.; Paus, T.; Barrett, J.; Dagher, A. Repetitive Transcranial Magnetic Stimulation of the Human Prefrontal Cortex Induces Dopamine Release in the Caudate Nucleus. *J. Neurosci.* **2001**, *21*, RC157. [CrossRef]
39. Miniussi, C.; Cappa, S.F.; Sandrini, M.; Rossini, P.M.; Rossi, S. The Causal Role of the Prefrontal Cortex in Episodic Memory as Demonstrated with RTMS. *Suppl. Clin. Neurophysiol.* **2003**, *56*, 312–320. [CrossRef]
40. Alcalá-Lozano, R.; Morelos-Santana, E.; Cortés-Sotres, J.F.; Garza-Villarreal, E.A.; Sosa-Ortiz, A.L.; González-Olvera, J.J. Similar Clinical Improvement and Maintenance after RTMS at 5 Hz Using a Simple vs. Complex Protocol in Alzheimer's Disease. *Brain Stimul.* **2018**, *11*, 625–627. [CrossRef]
41. Bagattini, C.; Zanni, M.; Barocco, F.; Caffarra, P.; Brignani, D.; Miniussi, C.; Defanti, C.A. Enhancing Cognitive Training Effects in Alzheimer's Disease: RTMS as an Add-on Treatment. *Brain Stimul.* **2020**, *13*, 1655–1664. [CrossRef]
42. Koch, G.; Bonnì, S.; Pellicciari, M.C.; Casula, E.P.; Mancini, M.; Esposito, R.; Ponzo, V.; Picazio, S.; Di Lorenzo, F.; Serra, L.; et al. Transcranial Magnetic Stimulation of the Precuneus Enhances Memory and Neural Activity in Prodromal Alzheimer's Disease. *Neuroimage* **2018**, *169*, 302–311. [CrossRef] [PubMed]
43. Vannini, P.; Hedden, T.; Becker, J.A.; Sullivan, C.; Putcha, D.; Rentz, D.; Johnson, K.A.; Sperling, R.A. Age and Amyloid-Related Alterations in Default Network Habituation to Stimulus Repetition. *Neurobiol. Aging* **2012**, *33*, 1237–1252. [CrossRef] [PubMed]
44. Avirame, K.; Stehberg, J.; Todder, D. Benefits of Deep Transcranial Magnetic Stimulation in Alzheimer Disease: Case Series. *J. ECT* **2016**, *32*, 127–133. [CrossRef] [PubMed]
45. Anderkova, L.; Eliasova, I.; Marecek, R.; Janousova, E.; Rektorova, I. Distinct Pattern of Gray Matter Atrophy in Mild Alzheimer's Disease Impacts on Cognitive Outcomes of Noninvasive Brain Stimulation. *J. Alzheimer's Dis.* **2015**, *48*, 251–260. [CrossRef]
46. Zhao, J.; Li, Z.; Cong, Y.; Zhang, J.; Tan, M.; Zhang, H.; Geng, N.; Li, M.; Yu, W.; Shan, P. Repetitive Transcranial Magnetic Stimulation Improves Cognitive Function of Alzheimer's Disease Patients. *Oncotarget* **2017**, *8*, 33864–33871. [CrossRef]
47. Petersen, R.C.; Smith, G.E.; Waring, S.C.; Ivnik, R.J.; Tangalos, E.G.; Kokmen, E. Mild Cognitive Impairment: Clinical Characterization and Outcome. *Arch. Neurol.* **1999**, *56*, 303–308. [CrossRef]
48. Cotelli, M.; Calabria, M.; Manenti, R.; Rosini, S.; Maioli, C.; Zanetti, O.; Miniussi, C. Brain Stimulation Improves Associative Memory in an Individual with Amnestic Mild Cognitive Impairment. *Neurocase* **2012**, *18*, 217–223. [CrossRef]
49. Eliasova, I.; Anderkova, L.; Marecek, R.; Rektorova, I. Non-Invasive Brain Stimulation of the Right Inferior Frontal Gyrus May Improve Attention in Early Alzheimer's Disease: A Pilot Study. *J. Neurol. Sci.* **2014**, *346*, 318–322. [CrossRef]
50. Drumond Marra, H.L.; Myczkowski, M.L.; Maia Memória, C.; Arnaut, D.; Leite Ribeiro, P.; Sardinha Mansur, C.G.; Lancelote Alberto, R.; Boura Bellini, B.; Alves Fernandes da Silva, A.; Tortella, G.; et al. Transcranial Magnetic Stimulation to Address Mild Cognitive Impairment in the Elderly: A Randomized Controlled Study. *Behav. Neurol.* **2015**, *2015*, 287843. [CrossRef]
51. Padala, P.R.; Padala, K.P.; Lensing, S.Y.; Jackson, A.N.; Hunter, C.R.; Parkes, C.M.; Dennis, R.A.; Bopp, M.M.; Caceda, R.; Mennemeier, M.S.; et al. Repetitive Transcranial Magnetic Stimulation for Apathy in Mild Cognitive Impairment: A Double-Blind, Randomized, Sham-Controlled, Cross-over Pilot Study. *Psychiatry Res.* **2018**, *261*, 312–318. [CrossRef]
52. Ruthirakuhan, M.T.; Herrmann, N.; Abraham, E.H.; Chan, S.; Lanctôt, K.L. Pharmacological Interventions for Apathy in Alzheimer's Disease. *Cochrane Database Syst. Rev.* **2018**, *5*, CD012197. [CrossRef]
53. Cui, H.; Ren, R.; Lin, G.; Zou, Y.; Jiang, L.; Wei, Z.; Li, C.; Wang, G. Repetitive Transcranial Magnetic Stimulation Induced Hypoconnectivity Within the Default Mode Network Yields Cognitive Improvements in Amnestic Mild Cognitive Impairment: A Randomized Controlled Study. *J. Alzheimer's Dis.* **2019**, *69*, 1137–1151. [CrossRef] [PubMed]
54. Raichle, M.E. The Brain's Default Mode Network. *Annu. Rev. Neurosci.* **2015**, *38*, 433–447. [CrossRef] [PubMed]
55. Kim, S.H.; Han, H.J.; Ahn, H.M.; Kim, S.A.; Kim, S.E. Effects of Five Daily High-Frequency RTMS on Stroop Task Performance in Aging Individuals. *Neurosci. Res.* **2012**, *74*, 256–260. [CrossRef] [PubMed]
56. Cohen, R.A.; Marsiske, M.M.; Smith, G.E. Neuropsychology of Aging. *Handb. Clin.Neurol.* **2019**, *167*, 149–180. [CrossRef]
57. Solé-Padullés, C.; Bartrés-Faz, D.; Junqué, C.; Clemente, I.C.; Molinuevo, J.L.; Bargalló, N.; Sánchez-Aldeguer, J.; Bosch, B.; Falcón, C.; Valls-Solé, J. Repetitive Transcranial Magnetic Stimulation Effects on Brain Function and Cognition among Elders with Memory Dysfunction. A Randomized Sham-Controlled Study. *Cereb. Cortex* **2006**, *16*, 1487–1493. [CrossRef] [PubMed]

58. Antczak, J.; Kowalska, K.; Klimkowicz-Mrowiec, A.; Wach, B.; Kasprzyk, K.; Banach, M.; Rzeźnicka-Brzegowy, K.; Kubica, J.; Słowik, A. Repetitive Transcranial Magnetic Stimulation for the Treatment of Cognitive Impairment in Frontotemporal Dementia: An Open-Label Pilot Study. *Neuropsychiatr. Dis. Treat.* **2018**, *14*, 749–755. [CrossRef]
59. Di Stasio, F.; Suppa, A.; Berardelli, A. Frontotemporal Dementia: A Neurophysiological Study. *Aging* **2018**, *10*, 2547–2548. [CrossRef]
60. Bonner, M.F.; Ash, S.; Grossman, M. The New Classification of Primary Progressive Aphasia into Semantic, Logopenic, or Nonfluent/Agrammatic Variants. *Curr. Neurol. Neurosci. Rep.* **2010**, *10*, 484–490. [CrossRef] [PubMed]
61. Finocchiaro, C.; Maimone, M.; Brighina, F.; Piccoli, T.; Giglia, G.; Fierro, B. A Case Study of Primary Progressive Aphasia: Improvement on Verbs after RTMS Treatment. *Neurocase* **2006**, *12*, 317–321. [CrossRef]
62. Trebbastoni, A.; Raccah, R.; de Lena, C.; Zangen, A.; Inghilleri, M. Repetitive Deep Transcranial Magnetic Stimulation Improves Verbal Fluency and Written Language in a Patient with Primary Progressive Aphasia-Logopenic Variant (LPPA). *Brain Stimul.* **2013**, *6*, 545–553. [CrossRef] [PubMed]
63. Benussi, A.; Dell'Era, V.; Cosseddu, M.; Cantoni, V.; Cotelli, M.S.; Cotelli, M.; Manenti, R.; Benussi, L.; Brattini, C.; Alberici, A.; et al. Transcranial Stimulation in Frontotemporal Dementia: A Randomized, Double-blind, Sham-controlled Trial. *Alzheimers Dement.* **2020**, *6*, e12033. [CrossRef]
64. Zhang, X.-Q.; Li, L.; Huo, J.-T.; Cheng, M.; Li, L.-H. Effects of Repetitive Transcranial Magnetic Stimulation on Cognitive Function and Cholinergic Activity in the Rat Hippocampus after Vascular Dementia. *Neural Regen Res.* **2018**, *13*, 1384–1389. [CrossRef] [PubMed]
65. Veitia, J.F.P. Repetitive Transcranial Magnetic Stimulation in the Rehabilitation of Vascular Dementia. Report of 2 Cases. *Biomed. J. Sci. Tech. Res.* **2019**, *19*, 14043–14046. [CrossRef]
66. León Ruiz, M.; Rodríguez Sarasa, M.L.; Sanjuán Rodríguez, L.; Benito-León, J.; García-Albea Ristol, E.; Arce Arce, S. Current Evidence on Transcranial Magnetic Stimulation and Its Potential Usefulness in Post-Stroke Neurorehabilitation: Opening New Doors to the Treatment of Cerebrovascular Disease. *Neurologia* **2018**, *33*, 459–472. [CrossRef]
67. Wang, F.; Geng, X.; Tao, H.-Y.; Cheng, Y. The Restoration After Repetitive Transcranial Magnetic Stimulation Treatment on Cognitive Ability of Vascular Dementia Rats and Its Impacts on Synaptic Plasticity in Hippocampal CA1 Area. *J. Mol. Neurosci.* **2009**, *41*, 145–155. [CrossRef] [PubMed]
68. Rektorova, I.; Megova, S.; Bares, M.; Rektor, I. Cognitive Functioning after Repetitive Transcranial Magnetic Stimulation in Patients with Cerebrovascular Disease without Dementia: A Pilot Study of Seven Patients. *J.Neurol. Sci.* **2005**, *229–230*, 157–161. [CrossRef]
69. Sedlackova, S.; Rektorova, I.; Fanfrdlova, Z.; Rektor, I. Neurocognitive Effects of Repetitive Transcranial Magnetic Stimulation in Patients with Cerebrovascular Disease without Dementia. *J. Psychophysiol.* **2008**, *22*, 14–19. [CrossRef]
70. Morrin, H.; Fang, T.; Servant, D.; Aarsland, D.; Rajkumar, A.P. Systematic Review of the Efficacy of Non-Pharmacological Interventions in People with Lewy Body Dementia. *Int. Psychogeriatr.* **2018**, *30*, 395–407. [CrossRef]
71. Liang, X.; Liu, K.; Guo, L. Repetitive Transcranial Magnetic Stimulation (RTMS): A Possible Novel Therapeutic Approach to Dementia with Lewy Bodies. *Med. Hypotheses* **2010**, *74*, 877–879. [CrossRef]
72. Jiang, Y.; Guo, Z.; McClure, M.A.; He, L.; Mu, Q. Effect of RTMS on Parkinson's Cognitive Function: A Systematic Review and Meta-Analysis. *BMC Neurol.* **2020**, *20*, 377. [CrossRef] [PubMed]
73. Lefaucheur, J.-P.; Aleman, A.; Baeken, C.; Benninger, D.H.; Brunelin, J.; Di Lazzaro, V.; Filipović, S.R.; Grefkes, C.; Hasan, A.; Hummel, F.C.; et al. Evidence-Based Guidelines on the Therapeutic Use of Repetitive Transcranial Magnetic Stimulation (RTMS): An Update (2014–2018). *Clin. Neurophysiol.* **2020**, *131*, 474–528. [CrossRef] [PubMed]
74. Takahashi, S.; Mizukami, K.; Yasuno, F.; Asada, T. Depression Associated with Dementia with Lewy Bodies (DLB) and the Effect of Somatotherapy. *Psychogeriatrics* **2009**, *9*, 56–61. [CrossRef] [PubMed]
75. Khedr, E.M.; Mohamed, K.O.; Ali, A.M.; Hasan, A.M. The Effect of Repetitive Transcranial Magnetic Stimulation on Cognitive Impairment in Parkinson's Disease with Dementia: Pilot Study. *Restor Neurol. Neurosci.* **2020**, *38*, 55–66. [CrossRef] [PubMed]
76. Bhattacharjee, S.; Kashyap, R.; Abualait, T.; Annabel Chen, S.-H.; Yoo, W.-K.; Bashir, S. The Role of Primary Motor Cortex: More Than Movement Execution. *J. Mot. Behav.* **2021**, *53*, 258–274. [CrossRef] [PubMed]
77. Macdonald, R.; Barnes, K.; Hastings, C.; Mortiboys, H. Mitochondrial Abnormalities in Parkinson's Disease and Alzheimer's Disease: Can Mitochondria Be Targeted Therapeutically? *Biochem. Soc. Trans.* **2018**, *46*, 891–909. [CrossRef] [PubMed]
78. Mammana, S.; Fagone, P.; Cavalli, E.; Basile, M.S.; Petralia, M.C.; Nicoletti, F.; Bramanti, P.; Mazzon, E. The Role of Macrophages in Neuroinflammatory and Neurodegenerative Pathways of Alzheimer's Disease, Amyotrophic Lateral Sclerosis, and Multiple Sclerosis: Pathogenetic Cellular Effectors and Potential Therapeutic Targets. *Int. J. Mol. Sci.* **2018**, *19*, 831. [CrossRef]
79. Andravizou, A.; Siokas, V.; Artemiadis, A.; Bakirtzis, C.; Aloizou, A.-M.; Grigoriadis, N.; Kosmidis, M.H.; Nasios, G.; Messinis, L.; Hadjigeorgiou, G.; et al. Clinically Reliable Cognitive Decline in Relapsing Remitting Multiple Sclerosis: Is It the Tip of the Iceberg? *Neurol. Res.* **2020**, *42*, 575–586. [CrossRef]
80. Leung, I.H.K.; Walton, C.C.; Hallock, H.; Lewis, S.J.G.; Valenzuela, M.; Lampit, A. Cognitive Training in Parkinson Disease: A Systematic Review and Meta-Analysis. *Neurology* **2015**, *85*, 1843–1851. [CrossRef]
81. Goel, V. Hemispheric Asymmetry in the Prefrontal Cortex for Complex Cognition. *Handb. Clin. Neurol.* **2019**, *163*, 179–196. [CrossRef] [PubMed]

82. Kodounis, M.; Liampas, I.N.; Constantinidis, T.S.; Siokas, V.; Mentis, A.-F.A.; Aloizou, A.-M.; Xiromerisiou, G.; Zintzaras, E. Hadjigeorgiou, G.M.; Dardiotis, E. Assessment of the Reporting Quality of Double-Blind RCTs for Ischemic Stroke Based on the CONSORT Statement. *J. Neurol. Sci.* **2020**, *415*, 116938. [CrossRef]
83. Rikos, D.; Dardiotis, E.; Aloizou, A.-M.; Siokas, V.; Zintzaras, E.; Hadjigeorgiou, G.M. Reporting Quality of Randomized Controlled Trials in Restless Legs Syndrome Based on the CONSORT Statement. *Tremor Other Hyperkinet. Mov.* **2019**, *9*, 1–6 [CrossRef]

healthcare

MDPI

Article

A Greek Validation Study of the Multiple Sclerosis Work Difficulties Questionnaire-23

Christos Bakirtzis [1,*,†], Artemios Artemiadis [2,†], Elli Nteli [1], Marina Kleopatra Boziki [1], Maria-Valeria Karakasi [3], Cynthia Honan [4], Lambros Messinis [5], Grigorios Nasios [6], Efthimios Dardiotis [7] and Nikolaos Grigoriadis [1]

1 Multiple Sclerosis Center, Second Department of Neurology, Aristotle University of Thessaloniki, GR 54124 Thessaloniki, Greece; nteli.elli@gmail.com (E.N.); bozikim@auth.gr (M.K.B.); ngrigoriadis@auth.gr (N.G.)
2 Faculty of Medicine, University of Cyprus, Nicosia CY 2029, Cyprus; artemiadis.artemios@ucy.ac.cy
3 Third University Department of Psychiatry, AHEPA University General Hospital, GR 54124 Thessaloniki, Greece; valeria28289@hotmail.gr
4 School of Psychological Sciences, College of Health and Medicine, University of Tasmania, TAS 7250 Launceston, Australia; cynthia.honan@utas.edu.au
5 Neuropsychology Section, University Hospital of Patras, GR 26504 Patras, Greece; lmessinis@upatras.gr
6 Department of Speech and Language Therapy, University of Ioannina, GR 45110 Ioannina, Greece; nasios@uoi.gr
7 Department of Neurology, University of Thessaly, GR 41500 Larisa, Greece; edar@med.uth.gr
* Correspondence: cbakirtzis@auth.gr; Tel.: +30-2310-994-665
† Equally contributing first authors.

Citation: Bakirtzis, C.; Artemiadis, A.; Nteli, E.; Boziki, M.K.; Karakasi, M.-V.; Honan, C.; Messinis, L.; Nasios, G.; Dardiotis, E.; Grigoriadis, N. A Greek Validation Study of the Multiple Sclerosis Work Difficulties Questionnaire-23. *Healthcare* **2021**, *9*, 897. https://doi.org/10.3390/healthcare9070897

Academic Editor: Konrad Rejdak

Received: 25 May 2021
Accepted: 13 July 2021
Published: 15 July 2021

Publisher's Note: MDPI stays neutral with regard to jurisdictional claims in published maps and institutional affiliations.

Abstract: The Multiple Sclerosis Work Difficulties Questionnaire-23 (MSWDQ-23) is a self-report instrument developed to assess barriers faced by People with Multiple Sclerosis (PwMS) in the workplace. The aim of this study was to explore the psychometric properties of the Greek version of the MSWDQ-23. The study sample consisted of 196 PwMS, all currently working in part- or full-time jobs. Participants underwent clinical examination and cognitive screening with the Brief International Cognitive Assessment for Multiple Sclerosis (BICAMS) and completed self-report measures of fatigue, psychological functioning, and quality of life, along with the MSWDQ-23 questionnaire. Confirmatory Factor Analysis (CFA) was performed, and goodness-of-fit measures were used to evaluate construct validity. Convergent validity was checked by correlating MSWDQ-23 scores with study measures. Cronbach's alpha value was produced to assess internal consistency. CFA yielded a model with a fair fit confirming the three-factor structure of the instrument. Higher work difficulties were associated with higher Expanded Disability Status Scale (EDSS) scores, poorer cognitive function, more fatigue, stress, anxiety, and depression, and poorer health status, supporting the convergent validity of MSWDQ-23. Internal consistency (Cronbach's alpha = 0.94) and test–retest reliability (ICC = 0.996, 95%, CI = 0.990–0.998) were excellent. The Greek MSWDQ-23 can be considered a valid patient-reported outcome measure and can be used in interventions aiming to improve the vocational status of PwMS.

Keywords: multiple sclerosis; employment; patient-reported outcome; MSWDQ-23; validation

1. Introduction

Multiple Sclerosis (MS) is the most common demyelinating and neurodegenerative disease of the central nervous system in young adults [1]. Loss of productivity is common in People with MS (PwMS) and is strongly related to higher levels of physical disability [2,3], reduced subcortical and cortical gray matter volumes [4], cognitive impairment [5], and higher self-perceived fatigue, anxiety, and mood [6,7]. According to an international survey conducted in Europe with more than 13,000 participants, an estimated 50% of working-age PwMS are unemployed, although significant variability of employment rates is observed across countries [8]. According to the results of this survey, cognitive deficits and fatigue substantially impact productivity, even in PwMS with low levels of physical disability. In Greece, the prevalence of MS is currently estimated to be 188.9 per 100,000 inhabitants, totaling approximately 21,000 PwMS [9]. Data about employment rates of PwMS in Greece are however scarce. To our knowledge, only one study has investigated employment in PwMS in Greece. This study found that only 32% out of a sample of 200 PwMS were fully or partially employed [10], indicating that there may be a substantially higher rate of unemployment among PwMS in Greece.

Several international studies have demonstrated the positive effects of engaging in and maintaining employment in PwMS. These positive effects are grounded in the associated improvements with quality of life [11] and self-esteem [12], intellectual enrichment (which enhances cognitive reserve) [13], and avoidance of inactive lifestyles [14], seen in those who are employed. On the contrary, unemployment may lead to social isolation [15], financial dependence on caregivers and social support systems, and relatedly, an inability to cope with disease-related costs [8]. The comprehensive assessment of MS-related work difficulties may therefore enable the identification of the factors that affect the vocational status of PwMS and the likely risk of work withdrawal, as well as the subsequent development of customized vocational rehabilitation programs [16].

The Multiple Sclerosis Work Difficulties Questionnaire 23 (MSWDQ-23) [17] is a shortened version of a 50-item questionnaire that evaluates various domains of working difficulties, categorized into three key dimensions (physical, psychological/cognitive, and external barriers). This 23-item questionnaire has adequate psychometric properties, including good internal consistency and construct validity, and has been validated for use in several languages [18–21]. This questionnaire enables the detection of MS-related occupational difficulties and has been suggested to be predictive of poor employment outcomes [22,23]. The aim of this study was to validate and assess the psychometric properties of the Greek version of the MSWDQ-23. Common, well-established clinical and cognitive tools, as well as self-report questionnaires were used in this study to assist with this validation. The results of this study indicate that the Greek version of the MSWDQ-23 has sound psychometric properties and maintains the three-factor structure of the original English version.

2. Materials and Methods

2.1. Study Population

This noninterventional, cross-sectional study was conducted at the Multiple Sclerosis Center of the Aristotle University of Thessaloniki. The study sample consisted of 196 PwMS in full- or part-time employment, recruited from the outpatient clinic. Inclusion criteria were: (a) age $\geq$ 18 years; (b) MS diagnosis according to the 2017 revised McDonald criteria [24]; (c) being currently employed; (d) the ability to perform all tests and procedures; and (e) no history of any inflammatory event at least 3 months prior to participation. The study was performed according to the Declaration of Helsinki and was approved by the local ethics committee (4.291/4). All participants provided written informed consent prior to their participation.

2.2. Tests and Measures

All participants completed a clinical examination by a neurologist where Expanded Disability Status Scale (EDSS) scores were determined, as well as cognitive screening by an experienced neuropsychologist using the Greek version of the Brief International Cognitive Assessment for MS (BICAMS) battery [25], which includes three tests (Symbol Digit Modalities Test (SDMT) [26]; Greek Verbal Learning Test (GVLT) [27]; Brief Visuospatial Memory Test-Revised (BVMT-R) [28]). The BICAMS battery was administered according to the proposed guidelines [29]. Participants also completed the Modified Fatigue Impact Scale (MFIS) to assess self-reported fatigue [30], the Depression, Anxiety, Stress Scale-21 (DASS-21) [31] to assess mood, the Multiple Sclerosis Impact Scale-29 (MSIS-29) [32] to quantify the impact of MS on daily living, and the EuroQoL-5 Dimensions scale (EQ-5D) [33] to assess health-related quality of life. Finally, participants also completed the MSWDQ-23. Prior to use, permission to validate the MSWDQ-23 was obtained by the authors. The COSMIN Study Design checklist for patient-reported outcome instruments (version July 2019) [34] was followed in order to validate this instrument. The questionnaire was translated back and forth by two independent English and Greek bilingual researchers, with all translation discrepancies reviewed on a case-by-case basis until consensus was reached. A final draft version of the questionnaire was trialed in 12 patients (who also provided feedback on the instrument's readability and comprehension), after which final amendments were made and approved by the study team. Demographic and clinical characteristics and occupational data were also collected. All tests were administered in the same order to all participants, in a quiet room with no distractions. All participants completed the questionnaires without assistance. Finally, 25 randomly chosen participants were re-administered the MSWDQ-23 two weeks later to assess the test–retest reliability of the scale.

2.3. Statistical Analysis

Data were checked for deviations from normality by visual inspection of histograms and Q-Q plots. Confirmatory Factor Analysis (CFA) was performed to evaluate the construct validity of the MSWDQ-23 using SPSS v22.0 and AMOS software for Windows (Armonk, NY: IBM Corp). Goodness-of-fit measures (and the corresponding cut-offs showing a good fit) to assess the model fit were the normed χ^2 (χ^2/df) to overcome the effect of the sample size (cut-off ≤ 3), the Root-Mean-Squared Error of Approximation (RMSEA, cut-off ≤ 0.08, for the 90% CI lower bound ≤ 0.05 and for the 90% CI upper bound < 0.10), the Standardized Root-Mean-squared Residual (SRMR, cut-off ≤ 0.08), and the Comparative Fit Index (CFI, cut-off ≥ 0.9) [35,36]. Items with standardized regression weights or loadings of less than 0.3 were assessed with a view toward removing such items from the model if present. Modification indices for error covariances of conceptually linked items over the value of 10 were identified and accounted for in the final model. In the context of construct validity, we assessed the relationship between the MSWDQ-23 and age (Pearson's rho correlation) and sex, disease type, education, and work type (the Mann–Whitney U-test was performed for these categorical variables). The absence of floor or ceiling effects was determined by the percentage of scores with maximum or minimum values respectively being less than 15%. Convergent validity was examined by correlating MSWDQ-23 scores with disease status variables (disease duration and EDSS), cognitive test scores (i.e., SDMT, GVLT, BVMT-R), and questionnaire scores (MFIS, DASS-21, MSIS-29, and EQ-5D). Correlation coefficients of >0.7, 0.51–0.7, 0.31–0.5, and 0.1–0.3 were considered very large, large, moderate, and small, respectively [21,37]. Cronbach's alpha was produced to assess internal consistency. Cronbach's alpha was also calculated after omitting items one-by-one in each subscale to assess the influence of each item on the subscale's internal consistency. The Intraclass Correlation Coefficient (ICC) with 95% Confidence Intervals (95% CI) for two-way random-effects models was used to assess test–retest reliability. A significance level of <0.05 was used for all analyses.

3. Results

3.1. Sample Characteristics

The study sample consisted of 126/196 (64.3%) females. The mean age was 38.6 years (SD = 10.0, range = 19–66). In total, 173/196 (88.3%) had Relapsing–Remitting MS (RRMS), 14/196 (7.1%) had Secondary Progressive MS (SPMS), and 9/196 (4.6%) had Primary Progressive MS (PPMS). There were 47/196 (24%) participants who completed primary/secondary education, while the remainder (76%) were educated at the tertiary level. The majority of the participants (107/196, 54.6%) were in full-time employment (i.e., working 8 or more hours per day) with the remainder in part-time employment (89/196, 45.4%). There were no missing values regarding the MSWDQ-23 instrument. Demographic and disease characteristics of the participants are presented in Table 1.

Table 1. Demographic and disease characteristics of the study population.

n	196
Female (%)	126 (64.3)
Mean age (SD, range)	38.6 (10.0, 19–66)
Mean disease duration (SD, range)	7.3 (7.1, 0–36)
Mean education years (SD, range)	15.2 (3.3, 6–27)
Employment	
Full-time (%)	107 (54.6)
Part-time (%)	89 (45.4)
Type of MS	
RRMS (%)	173 (83.3)
SPMS (%)	14 (7.1)
PPMS (%)	9 (4.6)
Median EDSS (SD, range)	2.0 (1.5, 1–7)

MS: Multiple Sclerosis; RRMS: Relapsing–Remitting Multiple Sclerosis; SPMS: Secondary Progressive Multiple Sclerosis; PPMS: Primary Progressive Multiple Sclerosis; EDSS: Expanded Disability Status Scale.

3.2. Confirmatory Factor Analysis

All items had standardized regression weights over 0.3; thus, there was no need for item removal. After inspection of the modification indices, eight error terms within the three subscales were correlated. The final fit statistics of CFA were $\chi^2/df = 1.741$ ($p < 0.001$), RMSEA = 0.062 (90% lower bound = 0.051, 90% upper bound = 0.072), SRMR = 0.054, and CFI = 0.936, indicating a fair fit of the data (see Figure 1 for factor correlations and item loadings).

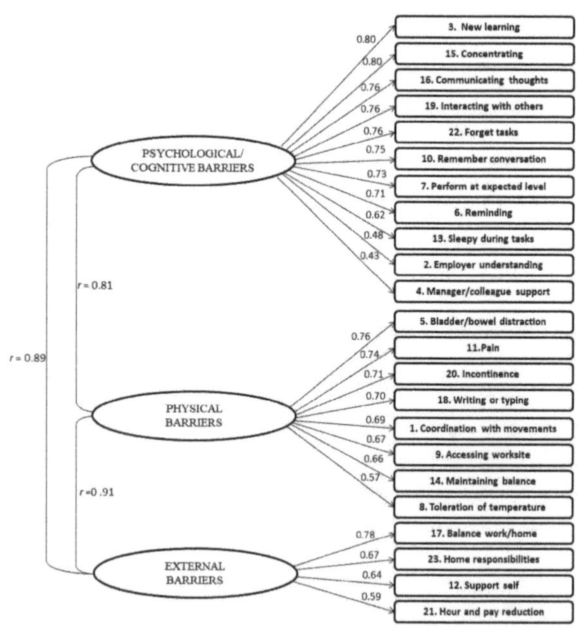

Figure 1. Factor structure of MSWDQ-23. The three latent factors (psychological/cognitive barriers, physical barriers, and external barriers) with their items (depicted in rectangles) and their standardized regression weights (i.e., factor loadings) are shown. The direction of the arrows represents the prediction of observed responses by the latent constructs.

3.3. Differences of Work Difficulties by Age, Gender, Education, Work Type, and MS Type

Age was positively correlated with psychological/cognitive barriers (rho = 0.199, $p = 0.005$), physical barriers (rho = 0.238, $p = 0.001$), external barriers (rho = 0.247, $p < 0.001$), and total MSWDQ-23 scores (rho = 0.245, $p = 0.001$). The effect of sex on MSWDQ-23 scores was nonsignificant (see Table 2). PwMS with a lower level of education (i.e., primary/secondary) had significantly more physical barriers in their work than those with higher education (i.e., tertiary). Furthermore, PwMS with more work barriers were significantly more likely to work part-time than full-time. Finally, participants with RRMS had significantly smaller MSWDQ-23 scores (i.e., less work barriers) than those with progressive MS (SPMS and PPMS). We considered the above findings as corroborative of the good construct and concurrent validity of the MSWDQ-23.

Table 2. Differences in MSWDQ-23 scores between patient groups. Values represent means ± standard deviations.

	Male	Female	p-Value⁋	Part-Time Work ¥	Full-Time Work ¥	p-Value⁋
Psychological-Cognitive Barriers	16.0 ± 16.6	18.2 ± 18.9	0.528	25.9 ± 20.0	10.0 ± 12.6	<0.001 *
Physical Barriers	17.7 ± 17.0	17.8 ± 21.2	0.360	27.2 ± 22.9	9.6 ± 11.7	<0.001 *
External Barriers	15.8 ± 20.4	21.5 ± 23.2	0.164	30.1 ± 23.8	10.1 ± 15.9	<0.001 *
Total MSWDQ-23	16.6 ± 15.5	18.6 ± 18.8	0.871	27.1 ± 19.4	9.9 ± 11.2	<0.001 *
	Primary/Secondary Education #	**Tertiary Education #**	**p-Value⁋**	**RRMS**	**Progressive MS**	**p-Value⁋**
Psychological-Cognitive Barriers	20.6 ± 18.9	16.4 ± 17.8	0.172	16.4 ± 17.7	24.8 ± 19.7	0.043 *
Physical Barriers	20.9 ± 18.0	16.8 ± 20.2	0.036 *	14.8 ± 17.9	40.5 ± 18.6	<0.001 *
External Barriers	20.9 ± 23.0	19.0 ± 22.2	0.711	18.0 ± 21.9	30.7 ± 23.0	0.008 *
Total MSWDQ-23	20.7 ± 17.8	17.0 ± 17.6	0.145	16.1 ± 16.9	30.3 ± 18.0	<0.001 *

Primary/Secondary education was defined as 12 or less years in education. Tertiary education was defined as more than 12 years in education; ¥ Part-time job was defined as less than 8 working hours per day. Full time job was defined as 8 or more working hours per day; ⁋ Mann-Whitney U tests; RRMS: Relapsing Remitting Multiple Sclerosis; MSWDQ-23: Multiple Sclerosis Work Difficulties Questionnaire-23; * $p \leq 0.05$.

3.4. Floor and Ceiling Effects

Amongst participants, 13.8% had a total MSWDQ-23 score of 0, and none had a score of 100. With regards to the three subscales, 21.4% had a score of 0 for psychological/cognitive barriers, 24.0% for physical barriers, and 33.7% for external barriers, indicating a significant floor effect for all subscales, after using the >15% cut-off. None of the participants had a score of 100 in any subscale. See Table 3 for the MSWDQ-23 descriptive statistics.

Table 3. Descriptive statistics of the MSWQ-23 subscales and total score.

	Mean	Median	Standard Deviation	Minimum	Maximum
Psychological-Cognitive Barriers	17.4	11.4	18.1	0	70.5
Physical Barriers	•17.8	12.5	19.8	0	90.6
External Barriers	19.5	12.5	22.4	0	87.5
Total MSWDQ-23	17.9	12	17.7	0	77.2

3.5. Convergent Validity

MSWDQ-23 subscales and total scores were significantly correlated with disease duration, disability, cognitive function, fatigue, MS-related psychological status, and overall health status (see Table 4). Scores showed small-to-moderate correlations with disease duration and EDSS, except physical barriers, which had a large correlation with EDSS scores, as expected. Small-to-moderate correlations were also present between the MSWDQ-23 and cognitive test scores. Moderate-to-large correlations were present between the MSWDQ-23 scores and DASS-21 subscale and EQ-5D scores. On the other hand, the MSWDQ-23 generally had large correlations with the MFIS and MSIS-29 scores.

Table 4. Convergent validity of the MSWDQ-23. Pearson's rho coefficients (p-values).

	Psychological-Cognitive Barriers	**Physical Barriers**	**External Barriers**	**Total MSWDQ-23**
Disease duration (years)	0.243 (0.001)	0.289 (<0.001)	0.198 (0.005) *	0.275 (<0.001)
EDSS	0.361 (<0.001) *	0.614 (<0.001) *	0.395 (<0.001)	0.503 (<0.001) *
SDMT	−0.345 (<0.001) *	−0.357 (<0.001) *	−0.238 (0.001) *	−0.360 (<0.001) *
CVLT-II	−0.245 (0.001) *	−0.244 (0.001) *	−0.141 (0.049) *	−0.247 (0.001) *
BVMT-R	−0.230 (0.001) *	−0.252 (<0.001) *	−0.157 (0.028) *	−0.245 (0.001) *
MFIS	0.778 (<0.001) *	0.752 (<0.001) *	0.712 (<0.001) *	0.831 (<0.001) *
Stress	0.580 (<0.001) *	0.505 (<0.001) *	0.548 (<0.001) *	0.601 (<0.001) *
Anxiety	0.521 (<0.001) *	0.508 (<0.001) *	0.477 (<0.001) *	0.558 (<0.001) *
Depression	0.585 (<0.001) *	0.525 (<0.001) *	0.493 (<0.001) *	0.600 (<0.001) *
MSIS-29	0.717 (<0.001) *	0.822 (<0.001) *	0.688 (<0.001) *	0.823 (<0.001) *
EQ-5D Mobility	0.423 (<0.001) *	0.587 (<0.001) *	0.341 (<0.001) *	0.511 (<0.001) *
EQ-5D Self-Care	0.328 (<0.001) *	0.490 (<0.001) *	0.311 (<0.001) *	0.421 (<0.001) *
EQ-5D Usual Activities	0.471 (<0.001) *	0.608 (<0.001) *	0.474 (<0.001) *	0.573 (<0.001) *
EQ-5D Pain/Discomfort	0.454 (<0.001) *	0.483 (<0.001) *	0.435 (<0.001) *	0.506 (<0.001) *
EQ-5D Anxiety/Depression	0.398 (<0.001) *	0.303 (<0.001) *	0.399 (<0.001) *	0.401 (<0.001) *
EQ-5D VAS	−0.373 (<0.001) *	−0.570 (<0.001) *	−0.452 (<0.001) *	−0.505 (<0.001) *

CVLT-II: California Verbal Learning Test-II; EDSS: Expanded Disability Status Scale; EQ-5D: EuroQuol-5D; MFIS: Modified Fatigue Impact Scale; MSIS-29: Multiple Sclerosis Impact Scale-29; MSWDQ-23: Multiple Sclerosis Work Difficulties Questionnaire 23; SDMT: Symbol Digit Modalities Test; VAS: Visual Analogue Scale; * $p \leq 0.05$.

3.6. Internal Consistency

The internal consistency of the overall MSWDQ-23 was excellent (see Table 5 for Cronbach's alpha values). Internal consistency for the subscales, according to the classification proposed by Kline [38], was excellent for psychological/cognitive barriers, good for physical barriers, and acceptable for external barriers, without any apparent unequal contribution of any one item. The ICC for psychological/cognitive (ICC = 0.991, 95% CI = 0.979–0.996), physical (IC = 0.989, 95% CI = 0.974–0.995), external (ICC = 0.986, 95% CI = 0.967–0.994), and total barriers (ICC = 0.996, 95% CI = 0.990–0.998) was excellent (see Table 5).

Table 5. Cronbach's alpha if item deleted for the MSWDQ-23 subscales.

Psychological-Cognitive Barriers	Cronbach's Alpha If Item Deleted	Physical Barriers	Cronbach's Alpha If Item Deleted	External Barriers	Cronbach's Alpha If Item Deleted
Item 2	0.910	Item 1	0.862	Item 12	0.726
Item 3	0.896	Item 5	0.859	Item 17	0.704
Item 4	0.911	Item 8	0.882	Item 21	0.718
Item 6	0.901	Item 9	0.860	Item 23	0.747
Item 7	0.901	Item 11	0.863		
Item 10	0.899	Item 14	0.858		
Item 13	0.906	Item 18	0.866		
Item 15	0.896	Item 20	0.860		
Item 16	0.899				
Item 19	0.900				
Item 22	0.897				
Total	0.910		0.879		0.778

4. Discussion

By utilizing CFA, this study verified the original three-factor structure of the Greek version of the MSWDQ-23. Furthermore, in accordance with previous validation studies, the Greek version of this instrument showed excellent internal consistency [17–21]. Importantly, there were no missing data, indicating that the Greek version of MSWDQ-23 is a highly feasible instrument. However, there were many participants with zero scores, and thus, there was high risk for a floor effect. It should be noted that the study sample consisted of PwMS currently working and with relatively mild disability (i.e., half of the participants had an EDSS score below 2.0). This rendered the participants less susceptible to work difficulties. Despite this finding, the MSWDQ-23 was significantly associated with other study measures (Tables 2 and 4), implying the presence of sufficient MSWDQ-23 score variance to produce meaningful associations and, as such, a reduced chance for a floor effect.

Indeed, with increasing age, more work barriers were reported, which is consistent with the expected physical, mental, and cognitive effects of prolonged disease duration, as well as aging [39,40]. Interestingly, the effect of sex was not significant, which was corroborated also by other similar studies [19,21]. This can be ascribed to the gender equality in Greece, such that men and women face similar workplace circumstances. As expected, PwMS with less education faced more physical barriers than those with higher education. Although not tested, we speculated that PwMS who have completed primary/secondary education would be more likely to have jobs requiring physical endurance than those who have completed higher tertiary education, thus explaining the role of physical barriers in PwMS with less education. In addition, this finding might suggest that a lower cognitive reserve reflects less capacity of the brain to adapt neuronally to changing demands due to MS-related disability [41,42]. Part-time workers reported more work barriers than full-time workers. This is consistent with the notion that PwMS, or even their employers, choose to limit their work demands in order to adjust for the high work difficulties [43]. In support of this, previous research has shown that PwMS reporting high work difficulties opt for larger reductions in work hours and may also change their type of work performed [17,20]. Finally, participants with RRMS reported fewer work difficulties than those with progressive MS, most probably signifying the different degree of disability between the two MS groups.

The MSWDQ-23 showed good convergent validity. This was indicated by its significant correlations with disease duration, disability, cognitive function, fatigue, psychological status, and MS-related and overall health status. Higher work difficulties were associated with higher EDSS, poorer cognitive function, more fatigue, stress, anxiety, and depression, and poorer quality of life and health status, a finding that was supported by the results of prior studies [17–21]. Since cognitive changes may predict the long-term clinical evolution of MS [44,45], it was not surprising that they were also associated with working difficulties. Notably, cognitive and psychological measures/barriers were associated with physical barriers/measures and vice versa (Table 4). This is in accordance with previous MS literature showing a strong interplay between work-related quality of life and physical, mental, and psychological health and quality of life in general [2,3,7].

This is the first study using standard cognitive testing (i.e., the BICAMS) and valid patient-reported outcomes (PROs) to explore the psychometric properties of the Greek version of the MSWDQ-23 questionnaire. However, some limitations should be noted. First, we did not use a healthy control group, which would help us ascribe the reported work difficulties to MS. From the available MSWDQ-23 validation studies, only one study used a healthy control group with good discrimination reported [21]; nevertheless, the hypothesis that the reported difficulties extend beyond that experienced by healthy individuals was not tested. Secondly, this study did not examine the future employment status of the participants, which would allow us to confirm the predictive validity of this instrument. However, previous studies have attested to the predictive validity of the instrument for future employment [21,22]. Thirdly, the study did not examine the concurrent validity of the MSWDQ-23 against other similar instruments, due to the absence of similar validated instruments for use in the Greek population. Finally, as mentioned before, half of our participants had mild disability, and this should be taken into account when generalizing the results of this study. Future research should focus on identifying the most important disease-related factors that negatively affect the working capacity of PwMS and on the ability of this instrument to elucidate them. In addition, further studies are needed to explore the predictive validity of the Greek MSWDQ-23 in relation to future changes in vocational status.

5. Conclusions

The Greek MSWDQ-23 demonstrated good psychometric properties in PwMS with mild and moderate disability. As such, the instrument can be considered a useful PRO for researchers and health professionals working with PwMS. The instrument also has the capacity to assist professionals working with PwMS and employers to gain valuable insight into the work difficulties faced by PwMS and to facilitate appropriate person-centered or tailored vocational amendments.

Author Contributions: Conceptualization, C.B., E.D. and N.G.; data curation, E.N. and G.N.; formal analysis, A.A.; investigation, C.B., E.N. and M.-V.K.; methodology, A.A.; project administration, N.G.; resources, C.B. and G.N.; supervision, C.H., E.D. and N.G.; validation, M.K.B., C.H. and L.M.; writing—original draft, C.B., A.A. and M.-V.K.; writing—review and editing, M.K.B., C.H., L.M. and N.G. All authors read and agreed to the published version of the manuscript.

Funding: This research received no external funding.

Institutional Review Board Statement: The study was conducted according to the guidelines of the Declaration of Helsinki and approved by the Ethics Committee of the Aristotle University of Thessaloniki, Faculty of Medicine (Protocol Code 4291; date of approval: 26 January 2021).

Informed Consent Statement: Informed consent was obtained from all subjects involved in the study. Written informed consent has been obtained from the patient(s) to publish this paper.

Data Availability Statement: The data presented in this study are available on reasonable request from the corresponding author. The data are not publicly available due to privacy reasons.

Acknowledgments: The authors would like to thank Eleni Grigoriadou and Theano Tatsi for providing administrative support.

Conflicts of Interest: The authors declare no conflict of interest.

References

1. Filippi, M.; Bar-Or, A.; Piehl, F.; Preziosa, P.; Solari, A.; Vukusic, S.; Rocca, M.A. Multiple sclerosis. *Nat. Rev. Dis. Primers* **2018**, *4*, 43. [CrossRef] [PubMed]
2. Krause, I.; Kern, S.; Horntrich, A.; Ziemssen, T. Employment status in multiple sclerosis: Impact of disease-specific and non-disease-specific factors. *Mult. Scler.* **2013**, *19*, 1792–1799. [CrossRef] [PubMed]
3. Glanz, B.I.; Dégano, I.R.; Rintell, D.J.; Chitnis, T.; Weiner, H.L.; Healy, B.C. Work productivity in relapsing remitting multiple sclerosis: Associations with disability, depression, fatigue, anxiety, cognition, and health-related quality of life. *Value Health* **2012**, *15*, 1029–1035. [CrossRef]
4. Jacobsen, C.; Zivadinov, R.; Myhr, K.M.; Dalaker, T.O.; Dalen, I.; Bergsland, N.; Farbu, E. Brain atrophy and employment in multiple sclerosis patients: A 10-year follow-up study. *Mult. Scler. J. Exp. Transl. Clin.* **2020**, *6*, 2055217320902481. [CrossRef]
5. Kavaliunas, A.; Danylaite Karrenbauer, V.; Gyllensten, H.; Manouchehrinia, A.; Glaser, A.; Olsson, T.; Alexanderson, K.; Hillert, J. Cognitive function is a major determinant of income among multiple sclerosis patients in Sweden acting independently from physical disability. *Mult. Scler.* **2019**, *25*, 104–112. [CrossRef]
6. Renner, A.; Baetge, S.J.; Filser, M.; Penner, I.K. Working ability in individuals with different disease courses of multiple sclerosis: Factors beyond physical impairment. *Mult. Scler. Relat. Disord.* **2020**, *46*, 102559. [CrossRef]
7. Raggi, A.; Covelli, V.; Schiavolin, S.; Scaratti, C.; Leonardi, M.; Willems, M. Work-related problems in multiple sclerosis: A literature review on its associates and determinants. *Disabil. Rehabil.* **2016**, *38*, 936–944. [CrossRef] [PubMed]
8. Kobelt, G.; Thompson, A.; Berg, J.; Gannedahl, M.; Eriksson, J.; MSCOI Study Group; European Multiple Sclerosis Platform. New insights into the burden and costs of multiple sclerosis in Europe. *Mult. Scler.* **2017**, *23*, 1123–1136. [CrossRef]
9. Bakirtzis, C.; Grigoriadou, E.; Boziki, M.K.; Kesidou, E.; Siafis, S.; Moysiadis, T.; Tsakona, D.; Thireos, E.; Nikolaidis, I.; Pourzitaki, C.; et al. The Administrative Prevalence of Multiple Sclerosis in Greece on the Basis of a Nationwide Prescription Database. *Front. Neurol.* **2020**, *11*, 1012. [CrossRef] [PubMed]
10. Yfantopoulos, J.; Grigoriadis, N.; Iliopoulos, I.; Karageorgiou, K.; Chantzaras, A.; Kyritsis, A.P.; Papathanasopoulos, P.; Tsimourtou, V.; Treska, X.; Tzortzis, D.; et al. Health and Economic Impact of Relapsing Forms of Multiple Sclerosis in Greece: The Storms Study. *Pharmacoeconomics* **2005**, *1*, 102.
11. Miller, A.; Dishon, S. Health-related quality of life in multiple sclerosis: The impact of disability, gender and employment status. *Qual. Life Res.* **2006**, *15*, 259–271. [CrossRef]
12. Bishop, M.; Timblin, R.I.; Rumrill, P.D. Medical, psychosocial, and vocational aspects of multiple sclerosis: Implications for rehabilitation professionals. *J. Rehabil.* **2016**, *82*, 6–13.
13. Sumowski, J.F.; Rocca, M.A.; Leavitt, V.M.; Riccitelli, G.; Comi, G.; DeLuca, J.; Filippi, M. Brain reserve and cognitive reserve in multiple sclerosis: What you've got and how you use it. *Neurology* **2013**, *80*, 2186–2193. [CrossRef]
14. Kahraman, T.; Savci, S.; Coskuner-Poyraz, E.; Ozakbas, S.; Idiman, E. Determinants of physical activity in minimally impaired people with multiple sclerosis. *Clin. Neurol. Neurosurg.* **2015**, *138*, 20–24. [CrossRef] [PubMed]
15. Johnson, K.L.; Yorkston, K.M.; Klasner, E.R.; Kuehn, C.M.; Johnson, E.; Amtmann, D. The cost and benefits of employment: A qualitative study of experiences of persons with multiple sclerosis. *Arch. Phys. Med. Rehabil.* **2004**, *85*, 201–209. [CrossRef]
16. Honan, C.A.; Brown, R.F.; Hine, D.W.; Vowels, L.; Wollin, J.A.; Simmons, R.D.; Pollard, J.D. The multiple sclerosis work difficulties questionnaire. *Mult. Scler.* **2012**, *18*, 871–880. [CrossRef] [PubMed]
17. Honan, C.A.; Brown, R.F.; Hine, D.W. The multiple sclerosis work difficulties questionnaire (MSWDQ): Development of a shortened scale. *Disabil. Rehabil.* **2014**, *36*, 635–641. [CrossRef] [PubMed]
18. Ellenberger, D.; Friede, T.; Sterz, C.; Flachenecker, P. Validation of the German version of the Multiple Sclerosis Work Difficulties Questionnaire (MSWDQ) [Abstract]. In Proceedings of the 31st Congress of the European Committee for Treatment and Research in Multiple Sclerosis, Barcelona, Spain, 7–10 October 2015; p. 128.
19. Martínez-Ginés, M.L.; García-Domínguez, J.M.; Forero, L.; Canal, N.; Rebollo, P.; Prefasi, D.; Honan, C.A.; Maurino, J. Spanish validation of a specific measure to assess work-related problems in people with multiple sclerosis: The Multiple Sclerosis Work Difficulties Questionnaire (MSWDQ-23). *Mult. Scler. Relat. Disord.* **2018**, *22*, 115–119. [CrossRef]
20. Kahraman, T.; Ozdogar, A.T.; Honan, C.A.; Ertekin, Ö.; Özakbaş, S. The Multiple Sclerosis Work Difficulties Questionnaire: Translation and cross-cultural adaptation to Turkish and assessment of validity and reliability. *Disabil. Rehabil.* **2019**, *41*, 2556–2562. [CrossRef]
21. Van Egmond, E.; van Gorp, D.; Honan, C.; Heerings, M.; Jongen, P.; van der Klink, J.; Reneman, M.; Beenakker, E.; Frequin, S.; de Gans, K.; et al. A Dutch validation study of the Multiple Sclerosis Work Difficulties Questionnaire in relapsing remitting multiple sclerosis. *Disabil. Rehabil.* **2019**, *8*, 1924–1933. [CrossRef]
22. Honan, C.A.; Brown, R.F.; Batchelor, J. Perceived cognitive difficulties and cognitive test performance as predictors of employment out-comes in people with multiple sclerosis. *J. Int. Neuropsychol. Soc.* **2015**, *21*, 156–168. [CrossRef]
23. Sterz, C.; Ellenberger, D.; Meißner, H.; Friede, T.; Flachenecker, P. Employment-associated factors in multiple sclerosis: Results from a cross-sectional study in Germany. *Edorium J. Disabil. Rehabil.* **2016**, *2*, 24–33. [CrossRef]
24. Thompson, A.J.; Banwell, B.L.; Barkhof, F.; Carroll, W.M.; Coetzee, T.; Comi, G.; Correale, J.; Fazekas, F.; Filippi, M.; Freedman, M.S.; et al. Diagnosis of multiple sclerosis: 2017 revisions of the McDonald criteria. *Lancet Neurol.* **2018**, *17*, 162–173. [CrossRef]

25. Polychroniadou, E.; Bakirtzis, C.; Langdon, D.; Lagoudaki, R.; Kesidou, E.; Theotokis, P.; Tsalikakis, D.; Poulatsidou, K.; Kyriazis, O.; Boziki, M.; et al. Validation of the Brief International Cognitive Assessment for Multiple Sclerosis (BICAMS) in Greek population with multiple sclerosis. *Mult. Scler. Relat. Disord.* **2016**, *9*, 68–72. [CrossRef]
26. Smith, A. *Symbol Digit Modalities Test: Manual*; Western Psychological Services: Los Angeles, CA, USA, 1982.
27. Vlachou, C.; Kosmidis, M.; Dardagani, A.; Tsotsi, S.; Giannakou, M.; Giazkoulidou, A.; Zervoudakis, E.; Pontikakis, N. Development of the Greek Verbal Learning Test: Reliability, Construct Validity, and Normative Standards. *Arch. Clin. Neuropsychol.* **2013**, *28*, 52–64. [CrossRef] [PubMed]
28. Benedict, R.H.B. *Brief Visuospatial Memory Test–Revised: Professional Manual*; Psychological Assessment Resources, Inc.: Odessa, FL, USA, 1997.
29. Langdon, D.W.; Amato, M.P.; Boringa, J.; Brochet, B.; Foley, F.; Fredrikson, S.; Hämäläinen, P.; Hartung, H.P.; Krupp, L.; Penner, I.K.; et al. Recommendations for a Brief International Cognitive Assessment for Multiple Sclerosis (BICAMS). *Mult. Scler.* **2012**, *18*, 891–898. [CrossRef] [PubMed]
30. Bakalidou, D.; Voumvourakis, K.; Tsourti, Z.; Papageorgiou, E.; Poulios, A.; Giannopoulos, S. Validity and reliability of the Greek version of the Modified Fatigue Impact Scale in multiple sclerosis patients. *Int. J. Rehabil. Res.* **2014**, *37*, 271–276. [CrossRef]
31. Lyrakos, G.N.; Arvaniti, C.; Smyrnioti, M.; Kostopanagiotou, G. Translation and validation study of the depression anxiety stress scale in the Greek general population and in a psychiatric patient's sample. *Eur. Psychiatry* **2011**, *26*, 1731. [CrossRef]
32. Hobart, J.; Lamping, D.; Fitzpatrick, R.; Riazi, A.; Thompson, A. The Multiple Sclerosis Impact Scale (MSIS-29): A new patient-based outcome measure. *Brain* **2001**, *124*, 962–973. [CrossRef]
33. Kontodimopoulos, N.; Pappa, E.; Niakas, D.; Yfantopoulos, J.; Dimitrakaki, C.; Tountas, Y. Validity of the EuroQoL (EQ-5D) instrument in a Greek general population. *Value Health* **2008**, *11*, 1162–1169. [CrossRef]
34. The COSMIN Study Design Checklist for Patient-Reported Outcome Measurement Instruments, Version July 2019. Available online: https://www.cosmin.nl/research-publications/ (accessed on 13 July 2021).
35. Kline, R.B. *Principles and Practice of Structural Equation Modeling*, 2nd ed.; Guilford Press: New York NY, USA, 2005.
36. Hu, L.; Bentler, P.M. Fit indices in covariance structure modeling: Sensitivity to under parameterized model misspecification. *Psychol. Methods* **1998**, *3*, 424–453. [CrossRef]
37. Hopkins, W.G. A New View of Statistics: Effect Magnitudes. 2002. Available online: www.sportsci.org/resource/stats/effectmag.html (accessed on 13 July 2021).
38. Kline, P. *The Handbook of Psychological Testing*, 2nd ed.; Routledge: London, UK, 1999.
39. Roy, S.; Frndak, S.; Drake, A.S.; Irwin, L.; Zivadinov, R.; Weinstock-Guttman, B.; Benedict, R.H. Differential effects of aging on motor and cognitive functioning in multiple sclerosis. *Mult. Scler.* **2017**, *23*, 1385–1393. [CrossRef] [PubMed]
40. Zeydan, B.; Kantarci, O.H. Impact of Age on Multiple Sclerosis Disease Activity and Progression. *Curr. Neurol. Neurosci. Rep.* **2020**, *20*, 24. [CrossRef]
41. Sumowski, J.F.; Leavitt, V.M. Cognitive reserve in multiple sclerosis. *Mult. Scler.* **2013**, *19*, 1122–1127. [CrossRef] [PubMed]
42. Nasios, G.; Bakirtzis, C.; Messinis, L. Cognitive Impairment and Brain Reorganization in MS: Underlying Mechanisms and the Role of Neurorehabilitation. *Front. Neurol.* **2020**, *11*, 147. [CrossRef] [PubMed]
43. O'Day, B. Barriers for People with Multiple Sclerosis Who Want to Work: A Qualitative Study. *J. Neurol. Rehabil.* **1998**, *12*, 139–146. [CrossRef]
44. Deloire, M.; Ruet, A.; Hamel, D.; Bonnet, M.; Brochet, B. Early cognitive impairment in multiple sclerosis predicts disability outcome several years later. *Mult. Scler.* **2010**, *16*, 581–587. [CrossRef]
45. Pitteri, M.; Romualdi, C.; Magliozzi, R.; Monaco, S.; Calabrese, M. Cognitive impairment predicts disability progression and cortical thinning in MS: An 8-year study. *Mult. Scler.* **2017**, *23*, 848–854. [CrossRef]

healthcare

MDPI

Systematic Review

Percutaneous versus Transcutaneous Electrical Stimulation of the Posterior Tibial Nerve in Idiopathic Overactive Bladder Syndrome with Urinary Incontinence in Adults: A Systematic Review

Aida Agost-González [1], Isabel Escobio-Prieto [1,*], Azahara M. Pareja-Leal [2], María Jesús Casuso-Holgado [1], María Blanco-Diaz [3] and Manuel Albornoz-Cabello [1]

[1] Departamento de Fisioterapia, Facultad de Enfermería, Fisioterapia y Podología, Universidad de Sevilla, 41009 Sevilla, Spain; aidaagost97@gmail.com (A.A.-G.); mcasuso@us.es (M.J.C.-H.); malbornoz@us.es (M.A.-C.)

[2] Servicio Madrileño de Salud, D.A. Norte, 28035 Madrid, Spain; azaharapl@gmail.com

[3] Departamento de Fisioterapia, Facultad de Medicina y Ciencias de la Salud, Universidad de Oviedo, 33006 Oviedo, Spain; blancomaria@uniovi.es

* Correspondence: iescobio@us.es

Citation: Agost-González, A.; Escobio-Prieto, I.; Pareja-Leal, A.M.; Casuso-Holgado, M.J.; Blanco-Diaz, M.; Albornoz-Cabello, M. Percutaneous versus Transcutaneous Electrical Stimulation of the Posterior Tibial Nerve in Idiopathic Overactive Bladder Syndrome with Urinary Incontinence in Adults: A Systematic Review. *Healthcare* **2021**, *9*, 879. https://doi.org/10.3390/healthcare9070879

Academic Editors: Grigorios Nasios, Lambros Messinis, Efthimios Dardiotis and Markos Sgantzos

Received: 6 June 2021
Accepted: 11 July 2021
Published: 13 July 2021

Publisher's Note: MDPI stays neutral with regard to jurisdictional claims in published maps and institutional affiliations.

Abstract: Background: Percutaneous electrical stimulation and transcutaneous electrical stimulation (PTNS and TTNS) of the posterior tibial nerve are internationally recognized treatment methods that offer advantages in terms of treating patients with overactive bladder (OAB) who present with urinary incontinence (UI). This article aims to analyze the scientific evidence for the treatment of OAB with UI in adults using PTNS versus TTNS procedures in the posterior tibial nerve. Methods: A systematic review was conducted, between February and May 2021 in the Web of Science and Scopus databases, in accordance with the PRISMA recommendations. Results: The research identified 259 studies, 130 of which were selected and analyzed, with only 19 used according to the inclusion requirements established. The greatest effectiveness, in reducing UI and in other parameters of daily voiding and quality of life, was obtained by combining both techniques with other treatments, pharmacological treatments, or exercise. Conclusions: TTNS has advantages over PTNS as it is more comfortable for the patient even though there is equality of both therapies in the outcome variables. More research studies are necessary in order to obtain clear scientific evidence.

Keywords: percutaneous electric nerve stimulation; transcutaneous electric nerve stimulation; adult; urinary bladder; overactive; urinary incontinence; tibial nerve

1. Introduction

The International Classification of Diseases (ICD-11) defines overactive bladder syndrome (OAB), with code GC50.0, as a urological condition characterized by voiding urgency, polyuria, and nocturia that may or may not be accompanied by urinary incontinence (UI) [1]. OAB presents a worldwide prevalence [2] of 16% to 23%, rising to 15% in those over the age of 40 years [3] and to 30–40% in those over 75 years [4], although it can be found in people of all ages. Its prevalence in Europe [5] is around 12%.

J.C. Angulo, in an article [6] published in 2016, revealed a 19.46% prevalence of OAB in the Spanish population, with at least one episode of urge UI (UUI) a day in 48.74% of cases [6]. This prevalence is greater in females than in males. Studies conducted [7] in the populations of Europe, the United States, Asia, and Africa reveal a prevalence of UUI of 1.5% to 14.3% in men aged between 18 and 20 years, whereas in women, it ranges from 1.6% to 22.8%. The same is true for those aged over 30 years, where the prevalence in men is from 1.7% to 13.3%, as opposed to 7% to 30.3% in women [7].

Among subjects with OAB on the global scale, UUI has been observed to be the most unpleasant symptom of this condition [4]. Further, people with OAB usually adopt certain coping strategies that involve a decrease in their quality of life and socialization, such as limiting liquid intake, avoiding traveling, and attempting to have direct access to toilets [3]

UI is the involuntary leakage of urine and lack of ability to control urination, accompanied by spontaneous contractions of the detrusor muscle. There are various subtypes of UI: urgency UI (UII), the sudden desire or need to urinate; stress UI (SUI), caused by efforts, physical exercise, sneezing, or coughing; mixed UI (MUI), combined with urgency and efforts [1] (code MF50.2 in ICD-11). To be able to determine the best treatment option in each patient, a personalized assessment is necessary, including the evaluation of different aspects of health, motivation, and availability or access to specific treatments [2].

Profitability is a fundamental aspect when it comes to reviewing treatment options in this type of condition, which entail great social and financial costs. Previous studies [7] showed a value of EUR 7 billion in subjects with OAB over 18 years old in Canada and European countries, including Spain [7].

There are several alternatives for OAB and UI treatment: behavioral treatments, considered first-line treatments; pharmacological or second-line treatments such as anticholinergic or antimuscarinic and b-adrenergic drugs, and, by way of a third line of treatment, injections of OnabotulinumtoxinA and therapies with electrical stimulation, including, among others, percutaneous and transcutaneous electrical stimulation (PTNS and TTNS, respectively), which are the object of this study [8].

With regard to treatment by electrical stimulation of the posterior tibial nerve (PTN), this involves retrograde stimulation of the nerve fibers of the sacral plexus, which innervates the bladder and detrusor muscle [2–5,8]. Electrical stimulation can be applied through insertion of a needle in the PTN—that is, PTNS is carried out in the said nerve—or through surface electrodes, with TTNS [9], with beneficial and safe effects in the short term in women with OAB, and no relevant adverse effects [10], according to the review by Sousa-Fraguas et al., 2020.

These techniques may represent an advantage in treatment of subjects with OAB who present UI, enabling these difficulties to be solved, as they can be compared favorably to treatment using antimuscarinic drugs, due to them being less costly [11].

In this respect, the present study aimed to summarize the knowledge available and conduct a critical analysis of the evidence from randomized controlled clinical trials, observational studies, systematic reviews, and meta-analyses on the effectiveness of PTNS and TTNS in the treatment of adults with OAB who present UI.

2. Materials and Methods

This systematic review followed the Preferred Reporting Items for Systematic Reviews and Meta-Analyses (PRISMA) guidelines [12]. The protocol was registered in the *International Prospective Register of Systematic Reviews* (PROSPERO/NHS)—number: 184809.

2.1. Selection Criteria

Three researchers independently reviewed the articles found. In order to formulate the objective and the question of the review, the PICOS strategy was used [13] (P—population or patients; I—intervention; C—comparison; O—outcomes; S—study design), in which P = (adults with OAB syndrome (OABS) and presence of UI); I = (PTNS and TTNS); C = (control group that received no intervention or received standard/usual care); O = (randomized clinical trials (RCTs), descriptive, observational studies, systematic reviews, and meta-analyses), and S = (randomized controlled clinical trials, descriptive observational studies, systematic reviews, and meta-analyses). This strategy enabled the establishment of critical reasoning on the issue [13] and the formulation of the following question: "What is the existing scientific evidence on the treatment of adults diagnosed as having OABS with UI through procedures of PTNS versus TTNS?".

2.2. Data Sources

The bibliographic search was performed between the months of February and May 2021. The search terms used were percutaneous electric nerve stimulation; transcutaneous electric nerve stimulation; adult; urinary bladder, overactive; urinary incontinence; tibial nerve. Two multidisciplinary databases, Scopus and Web of Science (WOS), were used in the search. The search strategy followed is presented in Table 1.

Table 1. Search strategy in WOS and Scopus databases.

Databases	Search Strategy
WOS (February–March 2021) (years 2015/2020)	I. ("Transcutaneous Electric Nerve Stimulation" OR "Therapy, Percutaneous Neuromodulation" OR "Electrical Neuromodulation, Percutaneous") AND adult AND ("Urinary Incontinence" OR "Urinary Bladder, Overactive") AND ("tibial nerve" OR "Posterior tibial nerve"). II. ("Transcutaneous electrical nerve stimulation" OR "Transcutaneous tibial nerve stimulation" OR "transcutaneous stimulation tibial nerve") AND adult AND ("overactive bladder" OR "detrusor activity" OR "urinary incontinence"). III. ("percutaneous tibial nerve stimulation" OR "PTNS") AND adult AND ("overactive bladder" OR "detrusor activity" OR "urinary incontinence").
Scopus (February–March 2021) (years 2015/2020)	I. ("Transcutaneous Electric Nerve Stimulation" OR "Therapy, Percutaneous Neuromodulation" OR "Electrical Neuromodulation, Percutaneous") AND adult AND ("Urinary Incontinence" OR "Urinary Bladder, Overactive") AND ("tibial nerve" OR "Posterior tibial nerve"). II. ("Transcutaneous electrical nerve stimulation" OR "Transcutaneous tibial nerve stimulation" OR "transcutaneous stimulation tibial nerve") AND adult AND ("overactive bladder" OR "detrusor activity" OR "urinary incontinence"). III. ("percutaneous tibial nerve stimulation" OR "PTNS") AND adult AND ("overactive bladder" OR "detrusor activity" OR "urinary incontinence").

2.3. Data Collection and Analysis

By way of exclusion criteria: all articles not published in English or Spanish; studies conducted in patients with neurological diseases or UI exclusively of neurogenic origin; carried out in children, animals, or patients with an associated underlying pathology; addressing fecal incontinence; in which treatment was not carried out with PTNS or TTNS of the PTN, or not aimed at treatment of OAB with UI; narrative or nonsystematic reviews; all documents not aligned with the research problem. The bibliographic research focused on all articles published from 2015 to 2020.

In order to obtain reliable, valid results, without them being influenced by bias, the Physiotherapy Evidence Database scale (PEDro) [14] was used to assess the methodological quality of the experimental studies, based on the Delphi list [15]. In the same way, the STROBE declaration [16] was applied for the evaluation of observational-type studies, and the PRISMA declaration [14] for reviews that followed its criteria in their execution. Articles that did not exceed the score of five in the PEDro scale [14] or with a score of less than 11 points in the STROBE declaration [16] were excluded, finally obtaining the articles chosen for the review.

3. Results

3.1. Literature Search

Figure 1 shows the PRISMA flow chart of this systematic review.

The initial search in the databases gathered a total of 259 articles, 98 from Web of Science (WOS) and 161 from Scopus.

The initial screening phase produced 130 articles after removing duplicates ($n = 129$).

Based on the titles and abstracts of the articles, a total of 56 articles were removed. Then, considering the remaining 74 eligible studies, many were excluded after full-text reading ($n = 47$), or because it was not possible to access the full text ($n = 3$), or for not passing the methodological quality scale ($n = 5$).

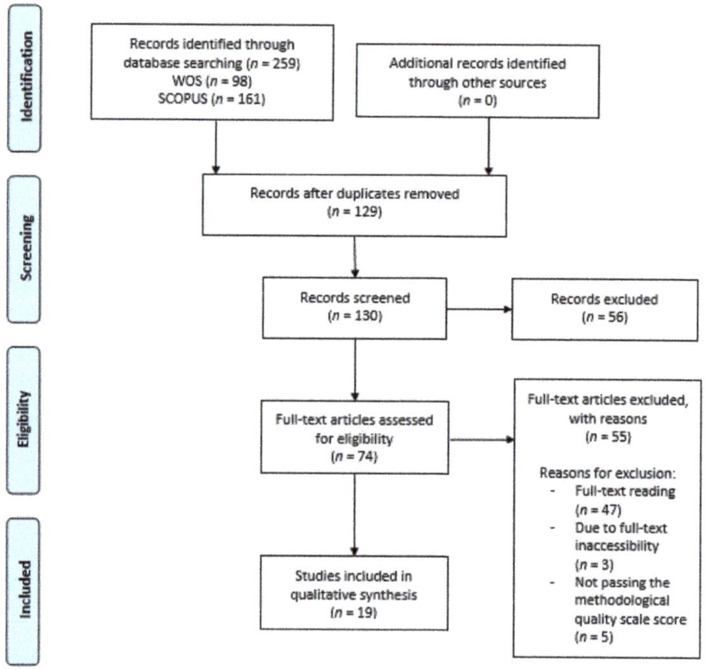

Figure 1. Flow chart of this systematic review.

Finally, 19 studies [3–5,17–32] were included. Of these, nine were experimental [3,17–24], including eight RTCs [3,17–22,24] (Table 2); four were observational studies [5,25–27] (Table 3), and six were either systematic reviews [4,29,31] or a meta-analysis [28], while two encompassed both types of studies [30,32] (Table 4).

Table 2. Characteristics of experimental studies included in the systematic review. Sevilla, ES. 2021.

Author/s	Study Design	Study Population	Sample Size	Intervention	Follow-Up	Randomization
Ramírez-García I., et al., 2019 [17]	RCT	OAB and OD	n = 68 (46 W and 22 M) (34 per G) Mean age (MA): 59.6 ± 27.4 kg. Symptom duration: 1–56 years with mean of 5.1.	Grupo A(GA) (intervention)→ TTNS Grupo B (GB) (control)→ PTNS	12 weeks	Online 1:1 Randomization Sequence"Sealed Envelope Lt. 2015"
Scaldazza C.V., et al., 2017 [18]	RCT	W with OAB	n = 60 W (30 per G) MA: 58.5 years (38–72)	GA→ EMS vaginal + exercises GB→ PTNS	1 m post-treatment.	Online randomization GraphPad QuickCalcs software
Preyer O., et al., 2015 [19]	RCT Pilot study	W with OAB, with no previous PTNS or anticholinergic treatment	n = 36 W (18 per G) GA: MA = 57.4 ± 9.5 GB: MA = 55.8 ± 16.2	GA→PTNS GB→ Tolterodine	3 m (baseline/1–3 m)	1:1 adapted randomization method (computer assistant)
Abulseoud A., et al., 2018 [20]	RCT	W with OAB and OD, and behavioral treatment failure	n = 30 W (15 per G) **G 1**: MA = 48 ± 16.42 **G 2**: MA = 48.13 ± 10.80	G 1: TTNS + placebo bottle identical to G 2. G 2: TTNS + Trospium Chloride	8 Weeks	1:1 by random number table
Martín-García M. & Cramptom J. 2019 [3]	RCT	W with non-neurogenic OAB who responded to PTNS initial treatment of 12 weeks	n = 24 W (12 per G) G PTNS: MA = 58 ± 10 G TTNS: MA = 54 ± 12	Maintenance G PTNS G TTNS	6 m Evaluation: baseline (post-treatment with 12 weeks PTNS), at 6 weeks, 3 m and 6 m	1:1 via sealed, opaque envelopes with numbered sequences
Seth J.H., et al., 2018 [21]	Randomized pilot trial	M/W with OAB after treatment with ineffective conservative therapy	n = 48 (24 multiple sclerosis and 24 Idiopathic OAB) G 1 (daily treatment): MA = 46.4(32–73), 18 W/6 M, 20 UI G 2 (weekly treatment): MA = 46.9(20–81), 20 W/4 M, 18 UI	TTNS G 1: 1 session/day G 2: 1 session/week	12 weeks, with evaluation at 4, 8 and 12 weeks	Stratified method by sealed envelopes
Welk B., et al., 2020 [22]	RCT	W with OAB, neurological diseases with urinary urgency with or without UI	n = 50. MA (73%, 22/30); 10% (3/30) used spontaneous voiding catheter G TTNS→ n = 26; MA = 62 (54–68) GS → n = 24; MA = 53 (46–64)	G TTNS→ TTNS. Increasing amplitude to maximum tolerance or flexion of the big toe GS→ TTNS. Constant amplitude	12 weeks	1:1 by random number generator
Tudor K.I., et al.020 [23]	Experimental study of two groups related retrospective	Neurological or idiopathic OAB refractory to 1st line treatments	n = 74 (52 W/22 M), MA = 56.0 (25.2, 59.8). 49 (66.2%)→ neurogenic OAB 25 (33.8%) → idiopathic OAB	PTNS + drug	12 weeks	Without randomization
Mallman S., et al., 2020 [24]	RCT	W with OAB	n = 50 (25 per G) MA = 61.48 ± 10.10	G transcutaneous sacral EMS G TTNS	6 weeks	Sequence generated in 2 G by WinPEPI version 11.63

Table 2. *Cont.*

Author/s	Variables	Results	Conclusions	Adverse Effects and Limitations	Met Quality
Ramírez-García I., et al., 2019 [17]	Difference day and night urination Mean voiding volume Number of urgencies and UI Quality of life Number of voiding in 24 h Number of UUI episodes	At the beginning/12 weeks. GA: $n = 34$/GB: $n = 34$ Differences by protocol: GA: $n = 32$ /GB: $n = 29$. Post intervention differences: GA: $n = 34$/GB: $n = 34$. Difference adjustment with 95% confidence interval Quality of life I-QoL: GA (21.5 p), GB (22.1 p) ($P < 0.001$) Number of daily voiding → GA→ $P = 0.0620$; GB→ $P = 0.0307$; Difference between groups (DG)→ $P = 0.3758$ Number of UUI episodes →GA: $P = 0.1293$; GB→ $P = 0.0009$; DG: $P = 0.0251$	Both techniques improve symptoms and quality of life Treatment adherence ($P = 0.236$).	No serious adverse effects Study carried out to measure differences between groups. High cost.	7/10
Scaldazza C.V., et al., 2017 [18]	Nocturia Quality of life (OAB-q SF) Bladder urgency perception (PPIU-S) Perception of global improvement (PGI-I)	Nocturia→ GA: $P = 0.1683$; GB→ $P = 0.0201$; DG: $P = 0.049$ Voiding volume→ GA: $P = 0.0048$; GB→ $P = 0.0003$; DG: $P = 0.0222$ OAB-q SF: 6 items: GA: $P = 0.0420$; GB: $P < 0.0001$; DG: $P = 0.0172$. 13 items: GA: $P = 0.0420$; GB: $P < 0.0001$; DG: $P = 0.0295$ PPIU-S: GA: $P = 0.1014$; GB: $P = 0.0001$; DG $P = 0.0459$ PGI-I→ GA: $P = 0.0415$; GB: $P = 0.0415$	PTNS 1st line treatment (efficacy, minimally invasive)	Without significant side effects	6/10
Preyer O., et al., 2015 [19]	Differences in number of voids in 24 h between both groups (voiding diaries). Number of UI episodes Quality of life (QoL-VAS)	$n = 16$ per group Number of voids→ without significant decrease 1–3 m in both G ($P = 0.13$), DGNS ($P = 0.96$). No significant differences in number at the beginning and post treatment ($P = 0.79$) Quality of life (QoL-VAS)→ depends on the initial values, in GB mean values are lower than GA in 1–3 m Increase in both G 1 and 3 m, without significant changes ($P = 0.07$) Number of UI episodes in 24 hours→ depends on the number episodes at the beginning of treatment ($P = 0.0001$). NSDG pre/post treatment ($P = 0.89$). Significant changes in 3 m (baseline/1 m) ($P = 0.03$)	Both are effective treatments: decrease in number of UI episodes, but not in urinary frequency. PTNS has fewer side effects.	The first 4 weeks mainly GB→ dry mouth and dizziness. 9 participants (3 m) GA→ pain in puncture area. 3 participants (3 m). Small sample size and standard deviations greater than expected, which could be due to a type II error; no blinding	6/10
Abulseoud A., et al., 2018 [20]	Brief OAB Symptom Score (OABSS) Short form of incontinence impact questionnaire (IIQ-7)	OABSS → post-treatment G ($P < 0.001$) and G 2 ($P = 0.024$) IIQ-7 → (G1: $P = 0.002$; G 2: $P = 0.001$). Pre- treatment min 50 points (good quality of life), post-treatment 20 (6 G 1 and 14 G 2). Pre- treatment→ severe OAB symptoms in 26 patients (12 G 1 and 14 G 2), post-treatment 4 (G 1). Cystometric volume → post-treatment (G1: $P = 0.026$; G 2: $P = 0.001$), G 2 ($P = 0.034$)	TTNS is tolerable and effective when combined with Trospium Chloride. Better results without side effects.	Without side effects Limitation: need for another group with sodium chloride treatment without TTNS. Longer follow-up studies needed.	8/10

Table 2. *Cont.*

Author/s	Variables	Results	Conclusions	Adverse Effects and Limitations	Met Quality
Martín-García M. & Cramptom J. 2019 [3]	Urinary frequency Number of urgency episodes Number of UUI episodes Severity of symptoms Quality of life (HRQoL)	Urinary frequency G TTNS→ decreasing from the beginning/6 m: 8.5 (1.9) vs 7.7 (2.8), (P = 0.373).; G PTNS→ during the study (P = 0.242), increasing from 7.3 (4.7) to 8.7 (2.4) at 6 m P = 0.208. Number of UUI episodes in 24 h: G TTNS→ (P = 0.900); G PTNS→ (P = 0.655) Number of urgency episodes in 24 h: G TTNS→ Friedman test (P = 0.038). Wilconxon test: significant increase baseline/ 6 Weeks: 1.7 (2.8) vs 3.2 (3.6), (P = 0.044); and baseline/3 m: 1.7 (2.8) vs 2.5 (2.6 (P = 0.011). Without significant changes baseline/6 m: 1.7 (2.8) vs 2.0 (1.4) (P = 0.325); G PTNS→ (P = 0.883) Severity of symptoms → G TTNS→ (P = 0.584); G PTNS→ (P = 0.854) Quality of life HRQoL→ G TTNS→ (P = 0.676); G PTNS→ (P = 0.948)	Application of bilateral TTNS is an effective and tolerable treatment for the maintenance of benefits in OAB symptoms with previous PTNS therapy For maintenance: PTNS, regular visit for consultation; Home TTNS	TTNS without side effects. PTNS → 3 minor episodes of needle insertion bleeding (2 participants); 1 episode of discomfort/ pain over the needle area	7/10
Seth J.H., et al., 2018 [21]	Quality of life (ICIQ-OAB) and (ICIQ-LUTqol) Part A for severity of symptoms and Part B for patient discomfort. 3-day voiding diary → urinary frequency in 24 h; number of UI episodes.	Quality of life → improvements in ICIQ-OAB score and ICIQ-LUTSqol score between the beginning and during 12 week treatments in both G. Daily treatment → part A - ICIQ-OAB → means improved between the beginning and 12 weeks from 9.3 (2.5) to 7.5 (3.1)Part B — ICIQ-OAB → from 29.6 (8.1) to 25.6 (9.5) Part A — ICIQ-LUTSqol→ from 51 (12.8) to 44.2 (13.1) Part B — ICIQ-LUTSqol → from 130.3 (43.7) to 105.5 (57.8) Weekly treatment→ part A - ICIQ-OAB→ week 12 from 9.1 (1.9) to 5.9 (1.7) Part B — ICIQ-OAB → from 29.7 (5.9) to 19.1 (8.5) Part A — ICIQ-LUTSqol → from 44.9 (9.0) to 35.9 (8.8) Part B — ICIQ-LUTSqol→ from 102.1 (40.1) to 63.9 (42.8) Urinary frequency in 24 h → Daily treatment → from 10.8 to 8.2 at 12 weeks Weekly treatment→ from 12.2 to 9.5 at 12 weeks Number of UI episodes → Daily treatment → from 2.8 to 1.6 at 12 weeks. Weekly treatment → from 2.3 to 0.9 at 12 weeks	Safe treatment. Low frequency stimulation (1 Hz) improves quality of life and symptoms of voiding diaries (daily/weekly) Neurological patients respond more frequently to treatment (65%) versus patients with idiopathic OAB (36%). No significant differences in tolerability between G	No significant complications during treatment or in satisfaction surveys 5 patients said therapy was uncomfortable and did not continue the treatment No significant safety problems. One patient developed skin redness in the area of stimulation Limitations: high dropout rate (in relation to device, lack of improvement, or local discomfort)	5/10

Table 2. *Cont.*

Author/s	Variables	Results	Conclusions	Adverse Effects and Limitations	Met Quality
Welk B., et al., 2020 [22]	Questionnaire (PPBC) Compress weight in 24 h, for UI 3-day voiding diary→ Urinary frequency and functional capacity in 24 h Quality of life (OAB-q SF, in G with OAB) Neurological patients → (NBSS), (Qualiveen-SF)	PPBC→ 13% (3/24) of sham patients and 15 (4/26) of TTNS treatment were considered responders (P = 0.77) Marginal mean of the end of the PPBC score was 3.3 (2.8–3.7) for TTNS vs 2.9 (2.5–3.4) for simulated (P = 0.30) Compress weight in 24 h→ NSDG (P = 0.64) Functional capacity → NSDG(P = 0.12) Urinary frequency in 24 h → NSDG(P = 0.32) OAB-qSF Questionnaire → NSDG in symptom discomfort (P = 0.82) and quality of life (P = 0.29) NBSS→ NSDG (P = 0.16) Qualiveen-SF→ NSDG (P = 0.85) Global assessment of improvement → NSDG (P = 0.27)	TTNS does not display greater efficacy in patient perception of OAB symptoms and objective parameters evaluated	With no adverse effects during the study Limitation in results generality, most had UI and had failed with pharmacological therapy. Small sample	8/10
Tudor K.I., et al., 2020 [23]	ICIQ-OAB questionnaire and ICIQ-LUTSqol 3-day voiding diary → urgency and severity of UI	64 (86%) completed 12 weeks. Significant improvements at 12 weeks of treatment in ICIQ-OAB ICIQ-LUTSqol, change in urinary frequency over 24 h and severity of UI in bladder diary G neurogenic VH→ ICIQ-OAB (P = 0.04); ICIQ-LUTSqol (P = 0.05) [in ICIQ-OAB, odds ratio (IC 95%) 0,93 (0,87, 0,99), P = 0,03], severity of UI [in bladder diary, odds ratio (IC 95%) 0.05 (0.01, 0.63), P = 0.02] and QoL [IUTQ-LUTSqol, odds ratio (IC 95%) 0.98 (0.96, 0.99), P = 0.007] at 12 weeks	PTNS is a possible alternative treatment in patients with neurological disease and with ineffective or intolerable 1st line treatment	No adverse effects. 5 patients had mild discomfort at the needle insertion area Lack of blinding, lack of a placebo or control group, and lack of urodynamic assessment before treatment Not validated questionnaires in patients with neurogenic OAB	5/10
Mallman S., et al., 2020 [24]	Quality of life: KHQ Severity of UI: ISI Discomfort due to OAB symptoms: OAB-V8	NSDG (P > 0.005) OAB-V8: (6 weeks P = 0.0019) G TPNS/G transcutaneous sacral EMS KHQ e ISI: NSDG	Both therapies are effective and safe for the treatment of women with OAB, UUI, and MUI	No side effects	6/10

W = women; M = men; OAB = Overactive bladder syndrome; OV = overactive detrusor; MA = mean age s; G = Group; PTNS = Percutaneous electrical stimulation of the posterior tibial nerve; TTNS = Transcutaneous electrical stimulation of the posterior tibial nerve; m = month; EMS = electrical stimulation; BMI = body mass index; UI = Urinary incontinence; PP = Per protocol; PIT = Per Intention to treat; DG = difference between groups; I- QoL = Urinary Incontinence Quality of Life Scale ;UUI = urgency urinary incontinence; OAB-qSF = Overactive Bladder Questionnaire Short Form; PPIU-S = Scale Patient Perception on Intensity of Urgency Scale; PGI-I = Patient Global Impressions Scale or Disease Improvement; QoL-VAS = Qol. Visual Analogue Scale; OABSS = Overactive Bladder Screening Scale; IIQ7 = Incontinence Impact Questionnaire-7; HRQoL = Health Related Quality of Life; ICIQ-OAB = International Consultation on Incontinence Questionnaire Overactive Bladder Module; ICIQ- LUT = International Consults on Incontinence Lower Urinary Tract Symptoms; ICIQ-LUTSqol = International Consults on Incontinence Lower Urinary Tract Symptoms Quality of Life; PPBC = Patient Perception of Bladder Condition; NBSS = Neurogenic Bladder Symptom Score; Qualiveen-SF = Short Form of Lower Urinary Tract Dysfunction Impact on Quality of Life; NSDG = non-significant difference between groups; IC = confidence interval

Table 3. Characteristics of the observational studies included in the systematic review. Sevilla, ES. 2021.

Author/s	Study Design	Study Population	*Sample Size*	Intervention	Follow-Up
Salatzki J., et al., 2019 [25]	Cross-sectional (cohort)	Positive response to PTNS treatments (10–12 weeks)	n = 83 PTNS-SEQ G 1: n = 28 G 2: n = 24 G 3: n = 31	G 1→ non-responders; no maintenance therapy G 2→ responders; possibility of maintenance therapy: they did not do it G 3→ responders who underwent maintenance therapy	18 weeks
Leroux P.A., et al., 2018 [5]	Prospective	Idiopathic or refractory OAB to anticholinergic treatment	n = 97 treated with TTNS 20 (21%) M; 77 (79%) W MA = 58.4 ± 16.6	**TTNS**	24 m
Moratalla-Charcos L.M., et al., 2018 [26]	Pilot study Prospective	OAB with or without OD/UI, without success in pharmacological treatment or dropout due to adverse effects	n = 45: 38 W and 7 M. MA = 66.6 ± 10.5 (41–83). OD: 53.3%.	**TTNS**	12 weeks
Palmer C., et al 2019 [27]	Retrospective	>65 with idiopathic OAB, after treatment with PTNS	n = 52: 23 M (44.3%); 29 W (55.8%). MA = 75.75 (65 to 93); BMI = 26.33 (17.4 to 43.9) kg/m^2	**PTNS**	12 weeks

Author/s	Variables	Results	Conclusions	Adverse Effects and Limitations	Met Quality
Salatzki J., et al., 2019 [25]	ICIQ-OAB→ UI in OAB ICIQ-LUT→ OAB symptoms. 3-day voiding diary PTNS-SEQ questionnaire→ variables observation to return to maintenance	Groups 2 and 3→ improvements compared to G 1. Patients with idiopathic or non-neurogenic OAB → significant improvement vs neurogenic OAB (P = 0.048). Group 3→ return to treatment after 39–204 days; significant improvements in nocturia (ICIQ-LUT, P = 0.036) and voiding diary (P = 0.046) To identify variables back to maintenance, (nocturia in 3 days/daytime urinary frequency/ number of UUI episodes), ICIQ-OAB, ICIQ-LUT→successful to distinguish between G 2 and 3 (Chi-squared 11.23, P = 0.047) PTNS-SEQ(P = 0.039)→ + 75% of cases. Increase in the categories "lack of treatment effect" — greater probability of belonging to G 2. Alternatives found to treatment of PTNS in PTNS-SEQ: (PTNS n = 28; PTNS home n = 20; PTNS in medicine clinic n = 20)	12 weeks of PTNS→ safe and effective treatment for OAB. A beneficial response with PTNS in nocturia was a factor to return to maintenance. The voiding diary offers more objective results for the evaluation of the treatment	No side effects Limitations: small sample size; difference in number of participants between G; results only applicable to public health.	21/22
Leroux P.A., et al., 2018 [5]	Questionnaire effectiveness USP and USP-OAB Treatment discontinuity and adverse effects Comorbidities Drugs during treatment follow-up	3 (3%) died of unknown cause/10 were lost at follow-up TTNS persistence and predictive factors → mean follow-up = 39.3 (25–65) m; mean persistence TTNS 8.3 (1–40 m). persistence = 12 m/28 patients (29%) e = 18 m/16 patients (16%) Discontinuity risk factors → At 3 m = 24 (28,9%) abandonment TTNS. Baseline score USP-OAB > 11 predictor of early treatment failure (P = 0.014) (univariate analysis)	TTNS treatment for refractory OAB. Few patients continued long-term therapy, probably due to a decrease in efficacy over time.	No adverse effects or pain Limitation: loss in follow-up during the study, and lack of placebo group, and objective urodynamic data	17/22

Table 3. *Cont.*

Author/s	Variables	Results	Conclusions	Adverse Effects and Limitations	Met Quality
Moratalla-Charcos L.M., et al., 2018 [26]	3-day voiding diary → Urinary frequency, nocturia, number of urgency episodes, number of UUI episodes, maximum voiding volume OABQ-SF Subjective improvements Satisfaction level	$n = 39/45 (86.6\%)$→ completed 12 weeks treatment Significant differences before and after treatment ($P < 0.05$) in urinary frequency, nocturia, number of urgency episodes, number of UUI episodes, maximum voiding volume. OABq-SF: $P > 0.05$. Statistically significant differences in GOD vs OAB Treatment satisfaction→ patients with mellitus diabetes ($P = 0.043$), in diabetes W ($P = 0.042$). In ordinal regression with 4 independent variables: number of vaginal deliveries ($P = 0.011$); psychiatric history ($P = 0.001$) were significant. Group with OD→ better satisfaction by increasing the number of vaginal deliveries and lower satisfaction for W and patients with diabetes. In ordinal regression with 3 independent variables: number of vaginal deliveries ($P = 0.05$) was significant	OAB treatment with TTNS is an effective, safe, minimally invasive and well tolerated therapy. In this study, all variables improved significantly compared to baseline.	Adverse effect: mild pain on plantar flexion, but no cases of dermatitis Limitation: lack of completion of voiding diaries, the OABQ-SF questionnaire, and subjective improvements, in addition to lack of a control group	17/22
Palmer C., et al 2019 [27]	OAB-V8 → OAB symptoms 3-day voiding diary Global impression of patient satisfaction (GIPS)	$n = 21$ (39%) used combination therapy during PTNS After PTNS→ 37 patients (70%) reported symptom improvements; 7 used anticholinergic, 6 used ß3 adrenoceptor agonist, 5 received intravesical injections of onabotulinumtoxnA, and 2 underwent sacral neuromodulation Mean old age→ $n = 13$, 1 or 2 medical comorbidities; $n = 10$, 3, or 4 medical comorbidities; $n = 6 + 5$. $n = 20$ W (69%) used anticholinergic treatments before PTNS; $n = 11$ W (38%) used combination therapy during PTNS	Effectiveness and viability of the PTNS technique for the treatment of OAB in elderly patients is observed, being able to choose as a 2nd line treatment. Decrease in the use of anticholinergics by PTNS therapy	Retrospective descriptive study, small sample size which could influence the results. More objective measures should have been used to determine the success of the treatment, such as voiding diary parameters or urodynamic parameters	13/22

PTNS = Percutaneous electrical stimulation of the posterior tibial nerve; PTNS-SEQ = Percutaneous Tibial Nerve Stimultion Service Evaluation Questionnaire; OAB = overactive bladder syndrome; TTNS = Transcutaneous electrical stimulation of the posterior tibial nerve; M = Man; W = Woman; UI = urinary incontinence; OD = overactive detrusor; G = group; m = month; BMI = body mass index; ICIQ-OAB = International Consultation on Incontinence Questionnaire Overactive Bladder Module; ICIQ-LUT = Lower urinary tract symptoms related quality of life questionnaire; UUI = urgency urinary incontinence; USP = Urinary Symptom Profile questionnaire; USP-OAB = Urinary Symptom profile- overactive bladder; OABQ-SF = Overactive Bladder Questionnaire Short Form; OAB-V8 = Overactive Bladder 8 questions Awareness Tool; GIPS = Global Impression of Patient Satisfaction; EMS = electrical stimulation; OAB-q = Short form Symptom Bother; OABSS = Overactive Bladder Symptom Score; UDI-6 = Urinary Distress Inventory, Short Form; IIQ-7 = Incontinence Impact Questionnaire; ICIQ-UI SF = International Consultation on Incontinence Questionnaire; LUTS = Lower Urinary tract Symptoms; SNM = sacral neuromodulation; QoL= Quality of life questionnaire.

Table 4. Characteristics of the systematic reviews and meta-analyses included in the systematic review. Sevilla, ES. 2021.

Author/s	Study Design	Number and Design of Studies	Study Participants	Inerventions	Variables	Results	Conclusions and Limitations
Wibison E., et al., 2015 [28]	Meta-analysis	16 studies 11 RCTs 5 non-comparative prospective studies	Participants with non-neurogenic OAB $n = 787$ 480; PTNS 108; antimuscarinics 63 combined therapy 136 sham or placebo treatment More W than M (10/16 W)	PTNS vs sham treatment PTNS vs antimuscarinics PTNS in non-comparative studies	Percentage of responders or patients with positive response Voiding diary parameters (urinary frequency; nocturia; UI and voiding volume)	PTNS vs sham procedure Urinary frequency and UI episodes PTNS vs antimuscarinics PTNS in comparative studies	PTNS therapy is effective for the short-term treatment of OAB, with greater efficacy than with sham treatment, and comparable with antimuscarinic drugs (but with fewer adverse effects). However, multimodal therapy was found to be more effective PTNS could be a maintenance therapy due to its safety and durability Dose, duration, frequency, pulse of PTNS, duration of study follow-up and demographic characteristics of the subjects were highly variable in the studies included
Booth J., et al., 2018 [4]	Systematic review	13 articles 10 RCTs 3 prospective studies	>18 years old with OAB with MUI $n = 629 \rightarrow 473$ (70%) W and 176 (28%) M, 16 (2%) gender is unknown 36 (18%) →sham treatment 142 (56%) →anticholinergic; 26 (10%)→ pelvic floor and bladder training 9 (4%) → sacral EMS or without treatment	Durability: 4–12 weeks (mean: 7.6 ± 3.6). Total number of sessions 5–90 (mean 21.6 ± 2.3) 30 min/individual session, except 3 of 20 min. 3 studies with daily stimulation, 7 studies with 2 times/week and 2 studies with 1 time/week	Urinary urgency symptoms Urinary frequency Nocturia Number of UI episodes Quality of life Adverse effects Urodynamic changes	Changes in voiding diary Score in OAB symptoms Effectiveness of TTNS Observational studies→ 3/3 General combined result→ 9/13	All studies observed improvements with TTNS treatment. It is safe and tolerable, due to this factor, its low cost, its ease of application and the possibility of self-administration by the patient, more studies are necessary to show its use as a 2nd line treatment
Tutolo M., et al., 2018 [29]	Systematic review	9 articles, all RCTs	Patients with OAB treated by SNM or PTNS	SNM and PTNS PTNS: 4 RCTs→ 388 patients PTNS vs Tolterodine → 3 m (94% W) PTNS vs sham therapy → 3 m PTNS vs placebo → 12 weeks PTNS vs vaginal electrical stimulation→ 12 weeks	Number of UI episodes and severity Number of compress in 24 h Urinary frequency Voiding volume Urinary urgency	PTNS efficacy PTNS safety	There are no high-quality studies able to guide professionals to choose between different treatments. This study shows that sacral stimulation and PTNS are safe and effective. SNM has more long-lasting effects, while PTNS needs to have maintenance treatment Limitations: number of results due to the impossibility of evaluator and patient blinding in the studies

Table 4. *Cont.*

Author/s	Study Design	Number and Design of Studies	Study Participants	Inerventions	Variables	Results	Conclusions and Limitations
Lo C.W., et al., 2020 [30]	Systematic review and meta-analysis	17 articles, all of them RCTs	Most patients with refractory OAB or patients who have tried 1st or 2nd line of treatment	Treatment with OnabotulinumtoxinA, SNM, and PTNS Heterogeneity between study designs, participants, follow-up, evaluated parameters, OnabotulinumtoxinA dose, and PTNS and SNM protocols	Quality of life Number of UUI episodes in 24 h Urinary frequency in 24 h improvement >50% of symptoms Nocturia Complications	Urinary frequency per day→ 9/17 Number of UI episodes per day→ 7/17 Improvement of 50% or more in symptoms at 12 weeks →8/17	The three modalities are effective and better than placebo for OAB treatment. At 12 weeks, SNM had greater efficacy in UI and urinary frequency, while OnabotulinumtoxinA had more complications Lack of enough data to conduct a meta-analysis in quality of life, urgency, UUI episodes/day, maximum bladder capacity and nocturia 4 studies rated as high risk of bias in the category of 'outcome measurement' because the self-report results could have been influenced by the placebo effect. Heterogeneity between studies made it difficult to clarify the results
Veeratterapillav R., et al., 2016 [31]	Systematic review	20 studies 6 RCTs 14 controlled clinical trials, prospective cohort studies, and retrospective series	Patients with OAB treated by PTNS and other comparative therapies.	PTNS vs pharmacological treatment PTNS vs placebo PTNS vs sham treatment Follow-up durability varied Different inclusion criteria and definitions of therapy "success" complicated the comparison	Urinary frequency Urinary urgency Nocturia UUI episodes Quality of life Urodynamic study	PTNS urodynamic results and PTNS clinical result PTNS vs anticholinergic therapy PTNS safety and other therapies PTNS cost	PTNS success changed due to informed symptoms by the patient (improvements in frequency and urgency), clinical evaluations (OAB and QoL questionnaires, voiding diary), and observation of aerodynamic variables. Hence, treatment was successful between 54.5% and 79.5% at 12 weeks and 15% and 71% in 1–3 years The studies suggested that PTNS efficacy is better with anticholinergic as a unique treatment, but there was limited evidence of combination therapy efficacy

Table 4. *Cont.*

Author/s	Study Design	Number and Design of Studies	Study Participants	Inerventions	Variables	Results	Conclusions and Limitations
Wang M., et al., 2020 [32]	Systematic review and meta-analysis	28 articles, 12 RCTs 16 observational studies	246 patients with OAB symptoms treated by PTNS and other comparative therapies	30 min PTNS for 12 weeks in 6/2 studies, the rest of studies had different protocols PTNS vs Tolterodine PTNS vs PTNS + sham treatment PTNS vs TTNS	3-day voiding diary (urinary frequency/day, nocturia/day, number of UUI episodes/day, number of UI episodes, daytime urination frequency/day, voiding volume and urodynamic data) Response rate	Urinary frequency/day → 10/28. Nocturia/día → 13/28 Number of UUI and UI episodes /day → 8/28 and 10/28 Urination frequency /day → 7/28 Voiding volume /day → 8/28 Urodynamic data Treatment response rate → 17/28 Comparison with other therapies	PTNS therapy was shown to be effective and safe for OAB treatment Limitations: heterogeneity of the studies included, however, a subgroup analysis was performed to observe that this factor was due to the study design. Second, evaluation of the improvements and success of the variable was done unconsciously Severe side effects: the most common was pain in the puncture area

RCT = randomized control trial; W = woman; M = man; OD = Overactive detrusor; MA = mean age; G = group; PTNS Percutaneous electrical stimulation of the posterior tibial nerve; TTNS = Transcutaneous electrical stimulation of the posterior tibial nerve; OAB = overactive bladder syndrome; UI = Urinary Incontinence; MUI = mixed urinary incontinence; UUI = Urgency urinary incontinence; EMS = electrical stimulation; SNM = sacral neuromodulation.

3.2. Summary of the Evidence

As for the comparison between PTNS and TTNS therapy, studies were found [3,17] in which significant changes were observed in the variables of diurnal frequency of urination, nocturnal frequency of urination, 24-h voiding frequency, mean voided volume, and number of episodes of UI and UUI in 24 h [17].

When TTNS combined with trospium chloride [20] was compared to placebo, a decrease in frequency of urination was observed in both groups ($p = 0.001$ and $p = 0.003$, respectively); as to mean voided volume, significant improvements were observed in both groups, with greater significance in the combined therapy ($p = 0.005$), although there was a significant delay in the combined therapy group with regard to the first sensation of a full bladder [20].

In other studies [23,32], upon combining PTNS with drugs, 35/53 participants completed the satisfaction survey after treatment, 66% of whom preferred to continue with maintenance treatment, with a mean interval of 44.4 days (7–155 days) and frequency of sessions of 1.1 months; attendance was observed if there were symptoms of OAB, while patients with multiple sclerosis had the possibility of returning [23].

Following this comparative line, one review was found [4] in which TTNS was compared with diverse therapies, including 3/10 studies that compared simulated therapy, 4/10 anticholinergic, 1/10 exercise, 1/10 behavioral, and 1/10 two different stimulation sites. The three remaining studies compared TTNS with other treatments: extended-release oxybutynin vs. TTNS + fármacoM; TTNS vs. transcutaneous sacral foramina vs. combination of the two; bladder and pelvic floor training vs. TTNS.

By contrasting daily or weekly treatment [21] with TTNS, 100% of weekly participants completed the compliance and experience questionnaire, in comparison to 90.5% of patients on daily therapy [21]. Although 53% (18) gave as a result a moderate or significant improvement in symptoms for the global response assessment (GRA), 75% (13/20) of neurological patients with OAB and 36% (5/14) of patients with idiopathic OAB responded to the intervention [21].

With respect to adherence to treatment [5,17] with TTNS, one of the studies [5] established different reasons for discontinuity (in 70 participants): lack of symptom relief (70%); difficulty in complying (6%); becoming asymptomatic (8%). However, 16.9% (14) of patients continued treatment, with a mean follow-up of 39.3 months [5].

Meanwhile, a BMI of obesity (=30 kg/m^2) was observed to be the only statistically significant variable predictive of failure in the response to PTNS ($p = 0.002$) [27]. Notably, after PTNS therapy, 66% (19/29) of participants informed of an improvement in their symptoms [27].

Some of the additional complications to those observed in the analysis of results [3,19,21,23,26,28,30,32] were urinary tract infections in 10/17 studies (peer comparisons revealed that OnabotulinumtoxinA was associated with a greater incidence of urine infections vs. placebo, sacral neurostimulation (SNS), and PTNS); ranking in order of fewest infections: first PTNS, second SNS, third placebo, and fourth OnabotulinumtoxinA. Further, urine retention with a need for intermittent catheterization was found in 11/17, with peer comparisons showing that OnabotulinumtoxinA was associated with a greater incidence of retention vs. placebo, SNS, and PTNS; ranking in order of lowest incidence: first SNM, second placebo, third PTNS, and fourth OnabotulinumtoxinA [30].

4. Discussion

The aim of this systematic review was to analyze the scientific evidence on the treatment of OAB with UI through procedures of PTNS, compared to TTNS, of the PTN. Nineteen studies were included, which analyze, observe, and compare these therapies with other methods, such as simulated treatment, placebo, anticholinergic or other drugs, sacral electrical stimulation, or vaginal electrical stimulation.

Among the studies whose intervention was based mainly on PTNS or TTNS therapy vs. another therapy, UI presented significant improvement when compared to placebo or simulated treatment [28,29,31]. Abulseoud A et al. [20] showed a significant improvement in the number of episodes of UI in combined groups of TTNS and trospium choloride compared to TTNS for eight weeks [20]. It is worth noting the significant improvements observed in the review by Veeratterapillay R et al. [31] in UUI after 12 weeks of treatment and two years of maintenance with PTNS therapy, unlike what was observed by Welk B et al. [22] in their RCT with PTNS therapy, with no significant differences between TTNS treatment compared to simulated therapy.

It is worth highlighting the improvements in UI reflected in the systematic review conducted by Booth J et al. [4], when combining TTNS therapy with pelvic floor exercises or behavioral treatment, as well as the results observed in the systematic review and meta-analysis performed by Wang M et al., as regards the reduction in the number of episodes of UI and UUI per day through PTNS therapy [32].

Apart from two studies analyzed [5,24], parameters related to frequency of urination, urgency of urination, and nocturia, as well as other symptoms of OAB such as voiding volume and urodynamic changes were included as variables and presented dissimilar results between studies.

In a recent study [10] from 2020, significant improvements were observed in the perception of quality of life of patients treated with TTNS and PTNS, with no differences between treatments [10].

By focusing on the quality of life observed in the studies analyzed, it is worth highlighting that all the experimental studies included provided data on this. Some of these studies [17,18,21,23] showed significant improvements in quality of life, through diverse questionnaires, after treatment with PTNS or TTNS, revealing that this improvement increased when TTNS was combined with trospium chloride, although the difference was not significant [20].

In 2013, Peters KM et al. [11] observed improvements in the quality of life of patients with OAB who were treated with PTNS, evaluated three years after treatment. In the present review, PTNS has been seen to present significant differences in quality of life when compared to vaginal electrostimulation [18], and it has been observed that there are significant differences in increased quality of life in both neurogenic and non-neurogenic OAB [23].

As for other parameters, it is worth noting the RCT of de Scaldazza CV et al. [18] in which significant differences were revealed in terms of the patient's global perception in favor of the PTNS technique compared to vaginal electrostimulation.

Leroux PA et al. [5] in their study showed some of the reasons why there is discontinuity in treatment with TTNS therapy, the most prevalent of which, in 70% of cases, was sufficient relief from symptoms, while in 6% it was due to complications for compliance with the treatment, and in 8% it was due to a complete reduction of the symptoms and becoming asymptomatic [5].

Most of the studies included in this review report of the absence of adverse effects during treatment [5,17,18,20,22,24,25]. Studies that combine PTNS and TTNS therapy [17], and those in which TTNS therapy is involved [5,20,22,25], point out that there are no adverse effects after the use of this therapy, except for the study conducted by Moratalla-Charcos LM et al. [26], who speak of mild pain on plantar flexion after the use of this technique.

Regarding PTNS therapy, no serious adverse effects were found, only minor bleeding episodes were mentioned or mild discomfort at the needle insertion site [3,25,29], sometimes causing hematomas or paresthesia at the puncture site [31].

As for the electrical stimulation parameters with PTNS therapy, most studies referred to weekly sessions for 12 weeks as the time of treatment. Although some of them did not show the other parameters, the rest coincided with regard to sessions of 30 min duration, a frequency of 20 Hz, pulse of 200 ms, 34-gauge needle inserted approximately five cm above the medial malleolus, and electrode in ipsilateral calcaneus [17,19,22,23,27]. In terms of amplitude, this was increased to the level of discomfort of the patient, feeling of tickling on the sole of the foot, or flexion of the big toe.

Upon referring to treatment with TTNS therapy, the parameters between the studies are more variable: some of the studies mentioned the same stimulation parameters as those of PTNS therapy, while others varied in terms of frequency, using a frequency of 10 Hz, as was the case of the randomized clinical trial of Welk B et al. [22], also highlighting the frequency of weekly sessions, with a total of three weekly sessions for 12 weeks.

The experimental studies of Abulseoud A et al. [20] and Seth JH et al. [21] stand out due to the use of different parameters, with the former [20] using frequencies of 10 Hz, pulse of 250 ms, treatment three times a week for eight weeks, and with a stimulation time of 30 min. Meanwhile, in the latter study [21], they used amplitudes of 27 mA, pulse between 70 and 560 ms, which varied depending on patient tolerability, for 12 weeks, both daily and weekly. Mallman S et al. [24], in their RCT, showed a stimulus duration of 20 min with a follow-up of six weeks and a pulse duration of 300 ms.

5. Conclusions

It is complicated to be able to establish which electrical stimulation therapy of the PTN is the most effective for treatments of idiopathic OAB with UI in adults, as far as the different parameters observed in this review are concerned, due to the variability of the results obtained and the electrical stimulation parameters used in the studies included. Nevertheless, it is worth highlighting the advantages TTNS therapy presents with respect to PTNS therapy, as this could be more comfortable for the patient, all things being equal in the results variable.

Author Contributions: Conceptualization, A.A.-G. and M.A.-C.; methodology, A.A.-G., M.A.-C., and I.E.-P.; formal analysis, A.A.-G., I.E.-P., and M.A.-C.; investigation, A.A.-G., A.M.P.-L., M.B.-D., and M.J.C.-H.; writing—original draft preparation, A.A.-G. and M.A.-C.; writing—review and editing, A.A.-G. and I.E.-P.; visualization, A.M.P.-L., M.B.-D., and M.J.C.-H. All authors have read and agreed to the published version of the manuscript.

Funding: This research received no external funding.

Institutional Review Board Statement: Not applicable.

Informed Consent Statement: Not applicable.

Conflicts of Interest: The authors declare no conflict of interest.

References

1. ICD-11-Mortality and Morbility Statistics. Available online: https://icd.who.int/browse11/l-m/en (accessed on 12 July 2021).
2. Kobashi, K.; Nitti, V.; Margolis, E.; Sand, P.; Siegel, S.; Khandwala, S.; Newman, D.; MacDiarmid, S.A.; Kan, F.; Michaud, E. A Prospective Study to Evaluate Efficacy Using the Nuro Percutaneous Tibial Neuromodulation System in Drug-Naïve Patients with Overactive Bladder Syndrome. *Urology* **2019**, *131*, 77–82. [CrossRef]
3. Martin-Garcia, M.; Crampton, J. A single-blind, randomized controlled trial to valuate the effectiveness of transcutaneous tibial nerve stimulation (TTNS) in Overactive Bladder symptoms in women responders to percutaneous tibial nerve stimulation (PTNS). *Physiotherapy* **2019**, *105*, 469–475. [CrossRef] [PubMed]
4. Booth, J.; Connelly, L.; Dickson, S.; Duncan, F.; Lawrence, M. The effectiveness of transcutaneous tibial nerve stimulation (TTNS) for adults with overactive bladder syndrome: A systematic review. *Neurol. Urodyn.* **2018**, *37*, 528–541. [CrossRef]
5. Leroux, P.A.; Brassart, E.; Lebdai, S.; Azzouzi, A.R.; Bigot, P.; Carrouget, J. Transcutaneous tibial nerve stimulation: 2 years follow-up outcomes in the management of anticholinergic refractory overactive bladder. *World J. Urol.* **2018**, *36*, 1455–1460. [CrossRef]
6. Angulo, J.C.; Brenes, F.J.; Lizarraga, I.; Rejas, J.; Trillo, S.; Ochayta, D.; Arumi, D. Impacto del número de episodios diarios de incontinencia de urgencia en los resultados descritos por el paciente con vejiga hiperactiva. *Actas Urol. Esp.* **2016**, *40*, 173–182. [CrossRef]
7. Milsom, I.; Coyne, K.S.; Nicholson, S.; Kvasz, M.; Chen, C.I.; Wein, A.J. Global prevalence and economic burden of urgency urinary incontinence: A systematic review. *Eur. Urol.* **2014**, *65*, 79–95. [CrossRef]
8. Raju, R.; Linder, B.J. Evaluation and treatment of Overactive Bladder in Women. In *Mayo Clinic Proceedings*; Elsevier LTD: Amsterdam, The Netherlands, 2020; Volume 95, pp. 370–377.

9. Valles-Antuña, C.; Pérez-Haro, M.L.; González-Ruiz de, L.C.; Quintás-Blanco, A.; Tamargo-Díaz, E.M.; García-Rodríguez, J.; San Martín-Blanco, A.; Fernandez-Gomez, J.M. Estimulación transcutánea del nervio tibial posterior en el tratamiento de la incontinecia urinaria de urgencia refractaria, de origen idiopático y neurogénico. *Actas Urol. Esp.* **2017**, *41*, 465–470. [CrossRef]
10. Sousa-Fraguas, M.C.; Lastra-Barreira, D.; Blanco-Díaz, M. Neuromodulación periférica en el síndrome de vejiga hiperactiva en mujeres: Una revisión). *Actas Urol. Esp.* **2020**, *45*, 177–187. [CrossRef]
11. Peters, K.M.; Carrico, D.J.; Wooldridge, L.S.; Miller, C.J.; MacDiarmid, S.A. Percutaneous tibial nerve stimulation for the long-term treatment of overactive bladder: 3-year results of the STEP study. *J. Urol.* **2013**, *189*, 2194–2201. [CrossRef]
12. Moher, D.; Liberati, A.; Tetzlaff, J.; Altman, D.G.; Grupo PRISMA. Preferred Reporting Items for Systematic Reviews and Meta-Analyses: The PRISMA Statement. *Rev. Esp. Nutr. Hum. Diet.* **2014**, *18*, 172–181. [CrossRef]
13. Methley, A.M.; Campbell, S.; Chew-Graham, C.; McNally, R.; Cherghi-Sohi, S. PICO, PICOS and SPIDER: A comparison study of specificity and sensitivity in three search tools for qualitative systematic reviews. *BMC Health Serv. Res.* **2014**, *14*, 579. [CrossRef]
14. Maher, C.G.; Sherrington, C.; Herbert, R.D.; Mosely, A.M.; Elkins, M. Reliability of the PEDro scale for rating quality of randomized controlled trials. *Phys. Ther.* **2003**, *83*, 713–721. [CrossRef]
15. Argimon Pallás, J.M.; Jimenez Villa, J. *Metodos de Investigación Clinica y Epidemiológica*, 3rd ed.; Elsevier: Barcelona, Spain, 2004.
16. Vandenbroucke, J.P.; von Elm, E.; Altman, D.G.; Gøtzsche, P.C.; Mulrow, C.D.; Pocock, S.J.; Poole, C.; Schlesselman, J.J.; Egger, M.; Strobe Initiative. Strengthening the Reporting of Observational Studies in Epidemiology (STROBE): Explanation and elaboration. *PLoS Med.* **2007**, *4*, 1628–1654. [CrossRef] [PubMed]
17. Ramirez-García, I.; Blanco-Ratto, L.; Kauffmann, S.; Carralero-Martínez, A.; Sánchez, E. Efficacy of transcutaneous stimulation of the posterior tibial nerve compared to percutaneous stimulation in idiopathic overactive bladder syndrome: Randomized control trial. *Neurourol. Urodyn.* **2019**, *38*, 261–268. [CrossRef]
18. Scaldazza, C.V.; Morosetti, C.; Giampieretti, R.; Lorenzetti, R.; Baroni, M. Percutaneous tibial nerve stimulation versus electrical stimulation with pelvic floor muscle training for overactive bladder syndrome in women: Results of a randomized controlled study. *Int. Braz. J. Urol.* **2017**, *43*, 121–126. [CrossRef] [PubMed]
19. Preyer, O.; Umek, W.; Laml, T.; Bjelic-Radisic, V.; Gabriel, B.; Mittlboeck, M.; Hanzal, E. Percutaneous tibial nerve stimulation versus tolterodina for overactive bladder in women: A randomized controlled trial. *Eur. J. Obstet. Gynecol. Reprod. Biol.* **2015**, *191*, 51–56. [CrossRef]
20. Abulseoud, A.; Moussa, A.; Abdelfattah, G.; Ibrain, I.; Saba, E.; Hassouna, M. Transcutaneous posterior tibial nerve electrostimulation with low dose trospium chloride: Could it be used as a second line treatment of overactive bladder in females. *Neurourol. Urodyn.* **2018**, *37*, 842–848. [CrossRef] [PubMed]
21. Seth, J.H.; Gonzales, G.; Haslam, C.; Pakzad, M.; Vashisht, A.; Sahai, A.; Knowles, C.; Tucker, A.; Panicker, J. Feasibility of using a novel non-invasive ambulatory tibial nerve stimulation device for the home-based treatment of overactive bladder symptoms. *Transl. Androl. Urol.* **2018**, *7*, 912–919. [CrossRef] [PubMed]
22. Welk, B.; Mckibbon, M. A randomized, controlled trial of transcutaneous tibial nerve stimulation to treat overactive bladder and neurogenic bladder patients. *Can. Urol. Assoc. J.* **2020**, *14*, 297–303. [CrossRef] [PubMed]
23. Tudor, K.I.; Seth, J.H.; Liechti, M.D.; Ochulor, J.; Gonzales, G.; Haslm, C.; Fox, Z.; Pakzad, M.; Panicker, J.N. Outcomes following percutaneous tibial nerve stimulation (PTNS) treatment for neurogenic and idiopathic overactive bladder. *Clin. Auton. Res.* **2020**, *30*, 61–67. [CrossRef] [PubMed]
24. Mallmann, S.; Ferla, L.; Rodrigues, M.P.; Paiva, L.L.; Sanches, P.R.S.; Ferreira, C.F.; Ramos, J.G.L. Comparison of parasacral transcutaneous electrical stimulation and transcutaneous posterior tibial nerve stimulation in women with overactive bladder syndrome: A randomized clinical trial. *Eur. J. Obstet. Gynecol. Reprod. Biol.* **2020**, *250*, 203–208. [CrossRef] [PubMed]
25. Salatzki, J.; Liechti, M.D.; Spanudakis, E.; Gonzalez, G.; Baldwin, J.; Haslam, C.; Pakzad, M.; Panicker, J.N. Factors influencing return for maintenance treatment with percutaneous tibial nerve stimulation for the management of the overactive bladder. *BJU Int.* **2019**, *123*, 20–28. [CrossRef] [PubMed]
26. Moratalla Charcos, L.M.; Planelles Gómez, J.; García Mora, B.; Santamaria Navarro, C.; Vidal Moreno, J.F. Efficacy and satisfaction with transcutaneous electrostimulation of the posterior tibial nerve in overactive bladder syndrome. *J. Clin. Urol.* **2018**, *11*, 331–338. [CrossRef]
27. Palmer, C.; Farhan, B.; Nguyen, N.; Ghoniem, G. Clinical outcomes of percutaneous tibial nerve stimulation in elderly patients with overactive bladder. *Arab. J. Urol.* **2019**, *17*, 10–13. [CrossRef] [PubMed]
28. Wibisono, E.; Rahardjo, H.E. Effectiveness of Short Term Percutaneous Tibial Nerve Stimulation for Non-neurogenic Overactive Bladder Syndrome in adults: A Meta-analysis. *Acta Med. Indones.* **2015**, *47*, 188–200.
29. Tutolo, M.; Ammirati, E.; van der Aa, F. What is New in Neuromodulation for Overactive Bladder? In *European Urology Focus*; Elsevier B.V.: Amsterdam, The Netherlands, 2018; Volume 4, pp. 49–53.
30. Lo, C.W.; Wu, M.Y.; Yang, S.S.D.; Jaw, F.S.; Chang, S.J. Comparing the efficacy of onabotulinumtoxinA, sacral neuromodulation, and peripheral tibial nerve stimulation as third line treatment for the management of overactive bladder symptoms in adults: Systematic review and network meta-analysis. *Toxins* **2020**, *12*, 128. [CrossRef]
31. Veeratterapillay, R.; Lavin, V.; Thorpe, A.; Harding, C. Posterior tibial nerve stimulation in adults with overactive bladder syndrome: A systematic review of the literature. *J. Clin. Urol.* **2016**, *9*, 120–127. [CrossRef]
32. Wang, M.; Jian, Z.; Ma, Y.; Jin, X.; Li, H.; Wang, K. Percutaneous tibial nerve stimulation for overacttive bladder syndrome: A systematic review and meta-analysis. *Int. Urogynecol. J.* **2020**, *31*, 2457–2471.

healthcare

MDPI

Review

Telerehabilitation of Post-Stroke Patients as a Therapeutic Solution in the Era of the Covid-19 Pandemic

Paulina Magdalena Ostrowska *, Maciej Śliwiński, Rafał Studnicki and Rita Hansdorfer-Korzon

Department of Physiotherapy, Medical University of Gdańsk, 7 Dębinki Street, 80-211 Gdańsk, Poland; sliwinskim@gumed.edu.pl (M.Ś.); rafal.studnicki@gumed.edu.pl (R.S.); rita.hansdorfer-korzon@gumed.edu.pl (R.H.-K.)
* Correspondence: paulina_graf@wp.pl

Abstract: (1) *Background*: Due to the pandemic caused by the SARS-CoV-2 virus, rehabilitation centres have become less available for neurological patients. This is the result of efforts to physically distance society, to try to slow the spread of the pathogen. Health care facilities were mainly restricted to urgent cases, while most physiotherapy treatments, mainly for patients with chronic conditions, were suspended. Some countries have seen a reduction in acute stroke hospital admissions of from 50% to 80%. One solution to the above problem is the use of telerehabilitation in the home environment as an alternative to inpatient rehabilitation. (2) *Aim of the study*: The purpose of this review is to analyse the benefits and limitations of teletherapy in relation to the functional condition of post-stroke patients. (3) *Methods*: Selected publications from 2019 to 2021 on the telerehabilitation of stroke patients were reviewed. The review was based on the Preferred Reporting Items for Systematic Reviews and Meta-Analysis (PRISMA) checklist. (4) *Results*: Studies have proven that teletherapy significantly improves the functional condition of post-stroke patients, resulting in improved quality of life and faster return to independence (while maintaining maximum possible precautions related to the SARS-CoV-2 virus pandemic). (5) *Conclusions*: Analysis of the study results showed comparable effectiveness of rehabilitation in the tele system to inpatient therapy. However, it should be emphasised that patients undergoing telerehabilitation must meet strict conditions to be eligible for this type of treatment program. However, the strength of the evidence itself supporting the effectiveness of this method ranks low due to the limited number of randomised control trials (RCT), small number of participants, and heterogeneous trials.

Keywords: Covid-19; telerehabilitation; post-stroke rehabilitation; virtual reality; stroke

Citation: Ostrowska, P.M.; Śliwiński, M.; Studnicki, R.; Hansdorfer-Korzon, R. Telerehabilitation of Post-Stroke Patients as a Therapeutic Solution in the Era of the Covid-19 Pandemic. *Healthcare* **2021**, *9*, 654. https://doi.org/10.3390/healthcare9060654

Academic Editor: Grigorios Nasios

Received: 9 May 2021
Accepted: 26 May 2021
Published: 31 May 2021

1. Introduction

Of the 15 million people worldwide who suffer a stroke each year, 5 million are permanently disabled and require ongoing care [1]. It is important to ensure access to rehabilitation within the first 3 months after stroke, as up to 80% of hospital readmissions are related to health habits and lack of appropriate therapy and post-stroke care [2]. Patients are generally discharged from hospital rehabilitation units 8–10 weeks after a stroke incident [3]. The functional condition of patients is characterised by significant motor deficits, and they require assistance with basic daily living activities (e.g., toilet use or eating) [4]. This is a period of increased post-damage (compensatory) plasticity in which major reorganizational changes occur in the brain. Therefore, it is critical to provide intensive, continuous, highly repetitive, task-oriented therapy to patients during this time. However, the reduction in the availability of inpatient post-stroke rehabilitation due to the SARS-CoV-2 virus pandemic has many serious consequences [5]. Above all, it puts patients at risk of acquiring pathological motor patterns and exacerbation of disability. Dhand's research shows that social isolation, which negatively affects the functional and emotional state of post-stroke patients, is an individual risk factor for stroke recurrence [6]. However, this single report should not be valid with respect to the entire population of stroke survivors. A potential solution to the problem of the increasing demand for rehabilitation services in resource-limited conditions is a teletherapy system, which involves the provision of rehabilitation services at home without direct contact between the therapist and the patient through communication technologies (mainly auditory, visual, tactile) [7,8]. Contact with the physiotherapist can be synchronous (communication with the therapist occurs over video chats during actual therapy time) or asynchronous (communication with the therapist occurs occasionally, outside of therapy time). Physiotherapists verify the progress, validate the correctness of motor tasks, and update therapy plans based on video conferencing feedback, system usage data, and game scores. The telerehabilitation system also uses virtual-reality-based therapy, which uses computer software to track users' movements and allows them to interact with a game or scenario presented on a television screen [9,10].

2. Materials and Methods

The review was based on research material obtained from PubMed, Directory of Open Access Journals, and Science Direct databases. The multi-search engine of the Main Library of the Medical University of Gdańsk was used to search for publications (search term: "telerehabilitation after stroke"). Study dates ranged from 2019 to 2021. Seventy-one records (PubMed—48, Directory of Open Access Journals—20, Science Direct—3) were identified and narrowed to the topics of "telerehabilitation" + "stroke". The results were restricted to full text only. After removing duplicates, the authors independently analysed the titles and abstracts of 28 papers. Ten publications were selected, evaluating the effectiveness of teletherapy with respect to various aspects of the functional status: motor skills of the directly involved upper limb, balance, cardiorespiratory fitness, cognitive function, and the level of acceptance and feasibility of teletherapy by its users at home. The selection was based on inclusion criteria: (1) study involving post-stroke patients, (2) describing therapeutic interventions using telerehabilitation, (3) determining the effect of teletherapy on the functional condition of post-stroke patients, (4) evaluating the effectiveness of teletherapy in relation to conventional rehabilitation, (5) written in English. The review was conducted in accordance with the PRISMA statement. (Figure 1) [11]. The question generated using the Population, Intervention, Comparison, Outcomes, Study Design (PICOS) components was: "In post-stroke patients (P), is teletherapy (I) an effective rehabilitation tool for improving the functional condition (O) in reference to inpatient rehabilitation (C)?" Only randomised controlled trials (RCT) were included in the qualitative synthesis because they are considered as the best research design for evaluating the effectiveness of clinical interventions (S) [12].

The review conducted according to the PRISMA method yielded 10 papers, which are shown in Table 1.

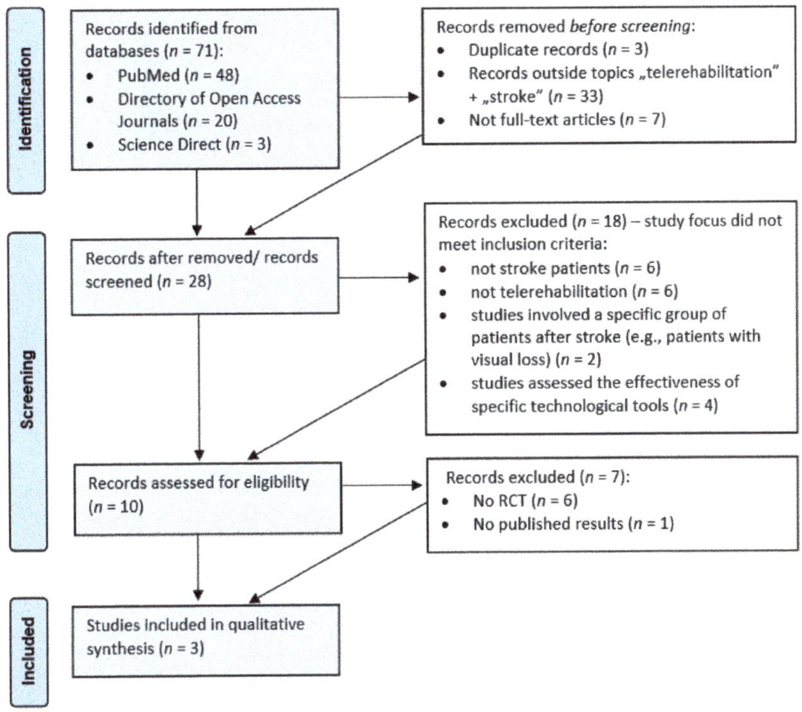

Figure 1. Flow diagram adapted from PRISMA which shows the process for identifying and screening the articles for inclusion and exclusion.

Table 1. Description of articles initially included by PRISMA methodology.

Article Type	Focus	Reference
RCT with published results	• Home-based TR for adults after stroke • Collaborative care model based TR for acute stroke patients • TR to improve balance in stroke patients	Cramer S. et al., 2019 Wu Z. et al., 2020 Burgos P. et al., 2020
RCT without published results	• Home-based VR rehabilitation after stroke	Sheehy L. et al., 2019
Case study	• User acceptance of a home-based stroke TR • Video home-based TR for stroke survivors	Chen Y. et al., 2020 Odetunde M. et al., 2020
Other trials (not randomized and/ or not controlled)	• Correlation of social network structure and home-based TR after stroke • TR aimed at increasing cardiorespiratory fitness for people after stroke • TR using to improve cognitive function in post-stroke survivors • Connecting stroke survivors and therapists using technology	Podury A. et al., 2021 Galloway M. et al., 2019 Torrisi M. et al., 2019 Simpson D. et al., 2020

3. Results

An analysis of the individual studies included in the review shows a high convergence in the inclusion and exclusion criteria for patients in the telerehabilitation group (Table 2) The average age of the participants was between 57 and 62 years old. Intervention sample sizes varied widely (e.g., in Cramer's study $n = 62$, in Burgos' study $n = 6$) [13,14].

Table 2. Inclusion and exclusion criteria for patients enrolled in the study program [15].

Inclusion Criteria:
1. Age ≥ 18 years
2. Past stroke (usually with time of stroke onset 4–36 weeks prior to study program)
3. Box and Block Test results of directly involved upper limb ≥ 3 blocks in 60 s
4. FM-A score of 22–56/66 points
5. Ability to transfer independently from sitting to standing
6. The presence of a caregiver while doing movement tasks
Exclusion Criteria:
1. Diagnosis (apart from the index stroke) significantly affecting the function of the upper or lower limb directly involved
2. A major, active, coexistent neurological or psychiatric disease (including dementia)
3. A medical disorder that substantially reduces the likelihood that a subject will be able to comply with all study procedures
4. Complete aphasia
5. Severe depression, defined as Geriatric Depression Scale Score > 10/15
6. Significant cognitive impairment, defined as Montreal Cognitive Assessment score < 22/30
7. Deficits in communication that interfere with reasonable study participation
8. Lacking visual acuity, with or without corrective lens, of 20/40 or better in at least one eye
9. Receipt of Botox to arms, legs or trunk in the preceding 6 months, or expectation that Botox will be administered to the arm, leg or trunk prior to study program
10. Unable or unwilling to perform study procedures/therapy, or expectation of noncompliance with study procedures/therapy
11. Pregnancy

A description of each publication, including information on the therapeutic interventions undertaken, the scales and tests used, and the results obtained, is provided below.

3.1. Impact of Virtual Reality (VR)-Based Telerehabilitation on the Functional Condition of Stroke Patients

In a randomised trial, Lisa Sheehy et al. evaluate the feasibility of virtual reality (VR)-based therapy at home as part of a telerehabilitation system for post-stroke patients [3]. In addition, qualitative and quantitative indicators of change in the functional status, ability to engage in the therapeutic programme, and the number and nature of adverse events are assessed. The therapy uses computer software that tracks the user's movements and allows them to interact with a game presented on a television screen. Twenty patients, who had survived a stroke in the past 18 months, and were able to maintain a standing position for at least 2 min, without cognitive impairment, were randomly assigned to the experimental and control groups. Patients in the experimental group used VR to train balance, correct gait stereotype, and reach with the directly affected upper limb. The control group used an iPad in therapy with training apps affecting cognitive ability, fine motor skills of the hand, and visual tracking. Both groups were instructed in the therapeutic programme prior to beginning therapy. The intensity and difficulty of the VR tasks were monitored and adjusted remotely. Therapy lasted 30 min, 5 days a week, for 6 weeks. The therapy used the Jintronix Rehabilitation system, designed to incorporate motor control principles. Patients were provided with a motion tracking camera (asynchronous user monitoring) and software to eliminate the need for gloves/controllers, etc. The patients underwent the following tests before and after completing the rehabilitation programme: Berg Balance Scale (BBS), Timed Up-and-Go (TUG), 5 Times Sit-to-Stand (FTSST), Community Balance and Mobility Scale

(CB&M), Stroke Impact Scale (SIS) (Table 3). Experimental group participants averaged 26.2 sessions, 27 min./session (77.8% compliance); control group participants averaged 33 sessions, 37 min./session (137.9% compliance) ($p = 0.11$ for sessions, $p = 0.002$ for the total duration). Mixed ANOVA showed no interaction between the group and time for any of the outcome measures. The only difference between groups concerned the FTSST test ($p = 0.017$). No falls or serious adverse events were reported [3,9,10].

Table 3. Scores of scales and functional tests before-post telerehabilitation [3,16].

Post-Pre Assessment Average (95% Confidence Intervals):	Experimental Group (EG)	Control Group (CG)	*p*-Values for Change Over Time (EG & CG Combined)
TUG (seconds)	−0.1 (−1.8, 1.6)	−1.4 (−3.7, 1)	0.326
TUG + cognitive task (seconds)	−1.7 (−4.2, 0.7)	−3.4 (−5.7, −1.1)	0.004
FTSST (seconds)	−3 (−5.8, −0.2)	−2 (−4.1, 0.2)	0.006
BBS (/56)	−0.5 (−4.2, 3.3)	0.6 (−0,5, 1.6)	0.959
CB&M (/96)	5.6 (−5, 16.2)	6.1 (1.6, 10.7)	0.049
SIS (/295)	7.7 (−2.1, 17.6)	13.8 (2.2, 25.3)	0.006

3.2. Effect of Telerehabilitation on Functional Condition of the Occupied Upper Limb and Level of Stroke Knowledge

In a study by Steven C. Cramer et al., conducted at 11 centres in the United States, 124 patients with upper limb motor deficits (Fugl–Meyer (FM) score 22–56), who were 4–36 weeks post-stroke, without cognitive impairment, were studied [14]. The effect of home-based telerehabilitation (combined with an educational module) after 36 treatment sessions (70 min each for 6 weeks) on the functional status of the directly affected upper limb and stroke knowledge was evaluated. The control group (IC) received therapy in an inpatient setting. Patients in the experimental group (TR) completed 35.4 of 36 (98.3%) assigned treatment sessions, while patients in the IC group completed 33.6 of 36 (93.3%) sessions. The mean change in FM score in the TR group increased by 7.86 points from baseline ($p < 001$), in the IC group by 8.36 points ($p < 001$). The covariance-adjusted mean change in FM score was 0.06 points (95% CI, −2.14 to 2.26) higher in the TR group ($p = 96$). The equivalence margin amounted to 2.47 and was outside the 95% CI, proving the comparable effectiveness of rehabilitation in both systems. At screening, patients in the TR group correctly answered an average of 22.4 of 30 questions (74.7%) from the stroke knowledge exam, whereas in the IC group, the number of correct answers averaged 22.8 of 30 questions (76%). After completion of the treatment programme, the outcome improved by 3.3 = 11% correct responses (TR group) and by 2.5 = 8.3% (IC group) [14,17,18].

3.3. Impact of Social Networks on the Course and Effects of Telerehabilitation

In contrast, Podury A. et al. examined the impact of social networks on the course of home telerehabilitation and the association of specific social factors with improved functional status and reduced depressive symptoms [2,4]. Thirteen patients who had experienced a stroke 2 to 16 months prior, with hand motor deficits, underwent supervised teletherapy. Key inclusion criteria: age $\geq$ 18 years, motor deficits of the upper limb directly involved in FM-A: 28–66 (if FM-A > 59, a Box and Block score needed to be obtained on the hemiparesis side > 25%, compared to the indirectly affected side, functional status of the upper limb scoring $\geq$ 3 blocks in 60 s); exclusion criteria: active neurological or psychiatric comorbidity, major depression (Geriatric Depression Scale score > 11), significant cognitive impairment (Montreal Cognitive Function Assessment score < 22), communication deficits [2,19]. The therapy programme lasted 12 weeks (1 h per day, 6 days per week). It included routine assessment of upper and lower limb motor function of the directly affected limb and mood. At the midpoint of the telerehabilitation programme (week 6), the researchers mapped each study participant's personal social network to assess the relationship between social network metrics and improvements in the functional status.

For this purpose, the personal network analysis and quantitative assessment tool for social network structure and composition, PERSNET, was used [6,20]. The results were compared with a historical cohort of 176 post-stroke patients (who did not receive telerehabilitation) to determine the differences in social networks. We demonstrated a correlation between network size and density and improved gait time ($p = 0.025$; $p = 0.003$). Social network density was associated with improved upper limb motor skills ($p = 0.003$), while network size was associated with reduced depressive symptoms ($p = 0.015$). Telerehabilitation patient networks were larger ($p = 0.012$) and less dense ($p = 0.046$) relative to historical patient networks. Median FM-A improved significantly from baseline 46 to 59 ($p = 0.0005$), median gait speed improved from a score of 0.94 to 1.01 ($p = 0.0007$), Geriatric Depression Scale decreased from 3 to 1 points ($p = 0.05$) [2,21,22].

3.4. *Effects of Tele Aerobic Training on Cardiorespiratory Fitness of Post-Stroke Patients*

Researchers Galloway M. et al. evaluated the feasibility and satisfaction levels of home aerobic training of 21 post-stroke patients (≥3 months after stroke) [23]. The exercises were held 3 days a week and were supervised by telephone. Post-stroke adults living with caregivers, able to walk independently (FAC score ≥ 3), and without cognitive impairment were eligible for the study. Participants underwent an 8-week programme of moderate to intense intensity (55–85% of maximum heart rate as determined by the Borg Rating of Perceived Exertion (RPE) between 13 and 16). Patients were assigned to one of four groups that differed in the length of the therapy session (1 group: 10 min./session, 2nd group: 15 min./session, 3rd and 4th group: 20 min./session). Prior to the commencement of telerehabilitation, each patient received instructions demonstrating the correct execution of each task of the therapeutic programme. Exercises were progressive, individually tailored to the patient's initial level of functional status, degree of disability, and movement preferences. Clinical assessment was used to modify motor tasks if the target heart rate was not reached or the patient experienced difficulty or discomfort. Cardiorespiratory fitness was measured by indirect spirometry during the first and last week of telerehabilitation during a 6-min walking test and a bicycle ergometer exercise test [24–27]. Patients' satisfaction with telerehabilitation was assessed using a questionnaire (23 multiple choice + 2 open-ended questions) at the end of the study. It was shown that 95% of the participants would undergo rehabilitation again in the tele system. On the one hand, the exercises turned out to be demanding enough to improve the efficiency and functional status of the patients; on the other hand, they were safe and convenient—the patients appreciated the therapy at home, which did not generate, for example, problems related to transportation [23,28–31].

3.5. *Assessing the Impact of Telerehabilitation Based on a Collaborative Model on the Functional Status of Post-Stroke Patients*

Chinese researchers have examined the effectiveness of a telerehabilitation programme for the purpose of improving post-stroke patients in the acute phase, based on a collaborative model [32]. Sixty-one subjects were studied (30 cases—intervention group, 31 cases—control group). Exclusion criteria included cognitive and psychiatric impairment, comorbidities affecting motor function, complete aphasia, and severe visual impairment. During hospitalisation, patients in the control group received routine early rehabilitation instructions. The content addressed physiological motor patterns of the upper and lower limbs, transfer from the lying position to a sitting one, or maintaining a proper range of motion in the joints. After hospital discharge, patients received rehabilitation and medication counselling by telephone (once a week). Patients in the intervention group received identical instructions during hospitalisation pertaining to early rehabilitation, but were provided remote home rehabilitation based on a collaborative model after hospital discharge. A care team consisting of neurologists, physiotherapists, nurses, counsellors, and caregivers was formed for this purpose. Remote in-home rehabilitation delivery was based on the TCMeeting v6.0 web-based videoconferencing system consisting of a computer, projector, camera, and data archiving software. The functional status of the patients was described by the Fugl–Meyer test, Berg Balance Scale, Up and Go Test, and a 6-Minute

Walking Test [32,33]. In addition, the Stroke Specific Quality of Life Scale (SSQoL) was used to assess the subjects' ability to perform daily living activities. Changes in functional condition and motor control ability were assessed at 4, 8, and 12 weeks after inclusion in the study (Table 4).

Table 4. Changes in scores of functional condition and motor control abilities after 4, 8 and 12 weeks of telerehabilitation [32].

	Intervention Group				Control Group				RM-ANOVA	
	Baseline	4 Weeks	8 Weeks	12 Weeks	Baseline	4 Weeks	8 Weeks	12 Weeks	F (Time*Group)	p
FM(UE)	11.93 ± 2.50	35.90 ± 2.78	49.10 ± 3.00	55.33 ± 2.81	2.61 ± 1.78	29.35 ± 2.36	39.35 ± 4.13	47.42 ± 3.90	42.523	<0.001
FM(LE)	13.37 ± 1.38	23.87 ± 1.28	25.50 ± 1.74	28.37 ± 2.51	14.13 ± 1.43	20.84 ± 1.39	24.23 ± 1.86	27.87 ± 1.73	57.000	<0.001
BBS	21.07 ± 3.29	30.50 ± 2.84	38.13 ± 2.84	43.13 ± 2.32	20.87 ± 2.33	28.06 ± 2.28	34.19 ± 2.15	38.29 ± 2.70	9.205	<0.001
TUG	41.93 ± 3.57	30.37 ± 3.62	22.73 ± 2.49	19.50 ± 2.73	40.58 ± 4.40	34.23 ± 2.86	27.13 ± 2.50	23.97 ± 3.35	16.320	<0.001
6MWT	91.73 ± 7.46	111.50 ± 8.12	128.90 ± 7.42	141.63 ± 8.68	92.35 ± 6.15	107.94 ± 5.14	123.13 ± 5.71	129.45 ± 7.06	10.530	<0.001

Note: F—time effect group (time factor * grouping factor).

3.6. Effect Of Telerehabilitation on Improving Cognitive Function in Post-Stroke Patients

Torrisi M. et al. examined the effect of telerehabilitation on improving cognitive function in post-stroke patients [34]. Forty patients were studied for this purpose (20 patients were assigned to the control group (CG), 20 to the experimental group (EG)). In the first phase of the programme, both groups underwent inpatient rehabilitation training: EG patients received cognitive training using VRRS-Evo (3D scenarios), CG patients underwent identical pencil and paper exercises. In the second phase of the programme (after hospital discharge), the EG group continued cognitive function therapy using a home VRR tablet (2D scenarios), whereas the CG group received traditional therapy (3 sessions per week, each session approximately 50 min.). The efficacy of telerehabilitation for the treatment of cognitive disorders was demonstrated, particularly improvements in global cognitive function levels, as well as in attentional, memory and language skills in EG [34,35].

3.7. Level of Acceptance of Telerehabilitation in Stroke Patients

Chen Yu et al. performed a qualitative study on the acceptance of home-based telerehabilitation by conducting in-depth interviews with 13 post-stroke patients (4–36 weeks after stroke), who completed 6 weeks of teletherapy [15]. Inclusion criteria included: age $\geq$ 18 years, FM-A score between 22–56, Box and Block test score of upper limb directly involved > 3 blocks in 60 s, no cognitive impairment. The telerehabilitation system consisted of four main components: games, exercise, stroke education, and telecommunications. The therapy programme consisted of 70-min sessions, 6 days a week, for 6 weeks. User acceptance of the telerehabilitation system was analysed using the Unified Theory of Acceptance and Use of Technology (UTAUT), a model of technology acceptance and use that describes four factors: expected outcomes, expected effort, facilitating conditions, and social impact [36]. The qualitative results of the study revealed high levels of participant satisfaction with teletherapy. Patients reported improvements not only in their functional status, but also in mood, cognitive ability, or social interaction. Telerehabilitation, on the one hand, provided repetitive tasks and, on the other hand, was characterised by a diversity of exercises. The system offered external (physiotherapist monitoring of outcomes) and internal (patient influenced progress in therapy) motivation to engage in the therapy programme. The flexibility of the time and location of the tele system sessions was an additional advantage, as it offset transportation issues. The only limitation, according to study participants, was barriers in technical/technological skills, but problems of this nature were resolved by contacting the research team [15].

3.8. Effect of Telerehabilitation on Balance Improvement in Post-Stroke Patients

Burgos P. et al. conducted a study evaluating the effect of telerehabilitation on balance improvement in post-stroke patients [13]. The tele system was based on games installed on smartphones. Additionally, participants were provided with inertial motion sensors (IMUs) and cloud databases. The research involved six patients 6–8 weeks after stroke Inclusion criterion: BBS score < 50, availability of caregiver during the patient's rehabilitation programme. Therapy lasted 4 weeks (a single session lasted 30 min.). The control group consisted of four people. Telerehabilitation was preceded by training the participants and their caregivers on safety and how to use the equipment. Balance was assessed with the Berg balance scale at the beginning and end of the rehabilitation programme [37,38]. In addition, the System Usability Scale (SUS) was used to evaluate the user experience. The scale consists of 10 questions relating to frequency and difficulty of use, system complexity, safety, or study participants' prior knowledge of the technology system. BBS scores improved with mean values: PRE = 35 ± 4.42 (62.5% ± 7.91), POST = 46.33 ± 3.01 (82.67% ± 5.37). Compared with the control group, the BBS PRE-POST variance in the study group was higher at 20.2% ± 6.36 vs. 12.5% ± 8.63, with the difference in variance between groups being statistically significant ($p = 0.019$). The mean SUS score was greater than 80 (87.5 ± 11.61), illustrating an excellent level of usability of the system for use [13,39–41].

3.9. *Feasibility of a Tele Therapeutic Program in Post-Stroke Patients*

Simpson D. et al. evaluated the degree of task feasibility in telerehabilitation, as well as the ability to verify exercise accuracy and functional progress in a tele system [42]. The study group consisted of 10 stroke survivors. Therapy lasted 4 weeks and included sitting and standing exercises [43]. The therapist remotely monitored adherence to movement commands and progress toward goals, and provided feedback to patients via the app. Inclusion criteria: age ≥ 18 years, stroke within the past two years, ability to stand up independently from a seated position from a chair. Each participant was instructed on proper task performance, safety and proper use of the technology. At baseline and at the end of the treatment programme, each patient completed a 2-min STS test and completed the System Usability Scale. Functional status was described using the Short Set of Physical Performance Battery (SPPB), which measures three aspects of physical status: gait speed, balance, and muscle strength. SPPB is a composite score from 0 to 12, where the higher the score, the better the patient's functional status. The change in scores from 0.99 to 1.34 points was considered clinically significant for older adults. During the study, participants performed a total of 224 exercise sessions out of the recommended 184 sessions. Individually, patients performed an average of 750 repetitions from sitting to standing (range 385–1410) over the course of 4 weeks, compared with an average of 724 prescribed repetitions (range 398–1395). The therapist made progress on the average goal from week 1 to week 4 by 92%. Participants rated the system usability at 79% [42,44,45].

3.10. *The Impact of Telerehabilitation on the Functional Condition of Stroke Survivors in African Countries*

Nigerian researchers, on the other hand, have developed a tele home exercise programme for post-stroke speakers of indigenous African languages [46]. The video-based program (VHEP) was conducted in the Yoruba language. In the course of its development, recommendations from the American Stroke Association were followed to include instruction in task-specific motor and postural exercises, trunk exercises, and correct gait stereotype exercises. Inclusion criteria: history of stroke three or more months ago, modified Ashworth Scale score ≥ 1, Brunnstrom score ≥ 3, no mental illness, epilepsy, cancer, heart disease. The research involved 10 people. The feasibility questionnaire was adapted from the Satisfaction Survey for the Individual with Stroke form from a smartphone educational intervention study. Each motor task was demonstrated on video for 5 min. The videos began with an introduction, followed by 5 short exercises in various positions (lying, kneeling, sitting, standing, walking). The main focus of the therapy was repetition, a gradual increase in task complexity, and functional training at a self-selected pace. The total time for a single exercise intervention was 30 min. Patients' exertion magnitude was monitored throughout the treatment programme based on feedback from the Modified Borg Perceived

Exertion Scale (RPE). The study proved that the VHEP system was goal-oriented and had a high level of acceptability for motor tasks. With a demonstration of each exercise, therapy proved easy and safe [18,46–48]. As an added bonus, the Yoruba language also made the tool accessible to patients who speak indigenous African languages [49].

Table 5 summarises publications based on randomised controlled trials (RCT) and their characteristics according to PRISMA methodology.

Table 5. Description of RCT included by PRISMA methodology in this review.

Authors/Year	Participants	Intervention	Outcomes Measurement	Results
Cramer S. et al., 2019	n = 124 (34 women, 90 men) (n = 62 in experimental group) Mean age (SD) of 61 years Adults ischemic stroke or intracerebral hemorrhage 4 to 36 weeks prior	Experimental and control group received 18 supervised and 18 unsupervised 70-min sessions. The treatment approach was based on an upper-extremity task-specific training manual and Accelerated Skill Acquisition Program All sessions for both groups included at least 15 min per day of arm exercises (the same 88 exercises for both groups) and at least 15 min per day of functional training. The TR system (for experimental group) consisted of an internet-enabled computer with table, chair, and 12 gaming input devices, but no keyboard, as no computer operation was required by patients.	FMA-UE (Fugl-Meyer Assessment upper extremity) Box and Block Test SIS (Stroke Impact Scale)	Both groups showed significant treatment-related motor gains, with a mean (SD) unadjusted FM score change from baseline to 30 days after therapy of 8.36 (7.04) points in the control group (p < 001) and 7.86 (6.68) points in the experimental group (p < 001). The adjusted mean change in FM score was 0.06 points larger in the experimental group (95% CI, −2.14 to 2.26; p = 96). The noninferiority margin (30% of the mean FM score change in the control group) was 2.47, which fell outside of this 95% CI, indicating that TR was not inferior to standard therapy on the primary end point [14]. Box and Block Test scores increased by 9.5 (p < 001) in the experimental group and by 8.8 (p < 001) in the control group and indicated noninferiority of TR therapy. Stroke Impact Scale hand motor domain scores increased by 23.7 (p < 001) in the experimental group and by 29.2 (p < 001) in the control group, although noninferiority was not demonstrated with this outcome [14].
Wu Z. et al., 2020	n = 61 (25 women, 36 men) (n = 30 in experimental group) Mean age (SD) of 58 years Adults ischemic or hemorrhagic stroke No information about time since stroke	Patients in the intervention group received home remote rehabilitation based on a collaborative care model. A collaborative care team consisting of neurologists, nurses, rehabilitation therapists, counselors and caregivers was established. Rehabilitation therapists assess the extent of patient dysfunction and work with family caregivers to develop rehabilitation plans and goals. The home remote rehabilitation guidance uses the Internet-based TCMeeting v6.0 video conferencing system. Patients in the control group received only routine rehabilitation and nursing measures, including dietary guidance, medication guidance and rehabilitation guidance, which were conducted by telephone follow-up once a week.	FMA-total (Fugl-Meyer Assessment total) FMA-UE (Fugl-Meyer Assessment upper extremity) FMA-LE (Fugl-Meyer Assessment lower extremity) BBS (Berg Balance Scale) TUG (Timed "Up&Go" Test) 6 MWT (6-min Walk Test) MBI (Modified Barthel Index)	See Table 4.

Table 5. *Cont.*

Authors/Year	Participants	Intervention	Outcomes Measurement	Results
Burgos P. et al., 2020	n = 10 (4 women, 6 men) (n = 6 in experimental group) Mean age (SD) of 61 years. Adults ischemic or hemorrhagic stroke in subacute (6–8 weeks after stroke)	Both groups received their standard rehabilitation treatment at the hospital site (3 sessions of 40 min per week of physical therapy for 4 weeks). In addition, the intervention group received 9 sessions of 30 min per week for 4 weeks. In each session, participants trained in balance tasks using smartphone-based exergames controlled by body motions.	BBS (Berg Balance Scale). MBT (Mini-BESTest). BI (Barthel Index) SUS (System Usability Scale)	Balance results improved in the BBS, with mean values of PRE = 35 ± 4.42 (62.50% ± 7.91), POST = 46.33 ± 3.01 (82.67% ± 5.37), and MBT PRE = 10.33 ± 2.87 (36.89% ± 10.26), and POST = 18.67 ± 2.81 (66.67% ± 10.01) with a statistically significant variation within PRE and POST ($F(1/5) = 60.84$, $p < 0.001$ and $F(1/5) = 45.96$, $p = 0.001$, respectively). Functional independence, measured by BI, also improved in the study group with PRE = 65.00 ± 4.47, and POST = 82.50 ± 8.80. There was also a statistically significant variation within PRE and POST times ($F(1/5) = 18.85$, $p = 0.007$) [13]. In comparison with the control participants, BBS variation PRE–POST for the study group was higher, with 20.20% ± 6.36 vs. 12.50% ± 8.63, with a statistically significant difference in the variation between groups ($F(1/7) = 9.15$, $p = 0.019$; Cohen-d = 2.98). For MBT PRE–POST variation, it was 29.7% ± 10.75 in the telerehabilitation group and 16.96% ± 9.39 in the control group, without significant differences between groups ($F(1/7) = 1.61$, $p = 0.245$; Cohen-d = 2.94). Functional independence (BI) in participants trained with our telerehabilitation system was higher compared to the controls: 17.50 ± 9.87 vs. 3.75 ± 8.53, with a significant difference between groups ($F(1/7) = 7.97$, $p = 0.025$; Cohen-d = 2.50) [13]. The average SUS score was higher than 80 (87.5 ± 11.61), which can be interpreted as an excellent system user usability level [13].

4. Discussion

This review examines the efficacy results of teletherapy on various aspects of the functional status of post-stroke patients. Research has shown that the tele system is a potential solution to the problem of increasing demand for rehabilitation services in a resource-constrained environment. Telerehabilitation provides continuity and an appropriate level of therapy intensity by using increased repetition of motor tasks [7,50]. Clinical research indicates that hundreds of repetitions in a specific movement pattern are required to achieve an optimal range of motor cortex neuroplasticity after stroke. Cramer's study calculated 1031 repetitions of upper limb movements per day in the intervention trial, demonstrating the effect of teletherapy in maximising the plasticity phenomenon by intensifying the applied therapy (the number of repetitions of upper limb movements during conventional therapy averages 32 per session) [14]. Telerehabilitation therefore contributes to a significant increase in function, regardless of whether therapy was initiated in the acute phase (<90 days from the stroke incident) or the chronic phase (>90 days after stroke). At the same time, it alleviates the problem of patient transport, which is cited as a major limitation in access to inpatient therapy, and reduces the expenses incurred for post-stroke rehabilitation in the private sector [34]. In turn, direct contact with a therapist, in the case of in-centre rehabilitation, plays an important role in the improvement process, increasing the patient's sense of security and support as well as the correctness of patients' motor tasks [42]. Despite this, in certain cases, rehabilitation in the tele system has comparable efficacy to therapy provided in a traditional inpatient environment, as evidenced by the results of the aforementioned review (Figure 2). Patients undergoing teletherapy are characterised by high commitment to the goals of the improvement programme (optimisation on the primary and secondary control scales) and satisfaction with rehabilitation outcomes (positive change in the Physical Activity Satisfaction Scale) [42]. The effectiveness of post-stroke therapy is closely related to the patient's motivation; however, maintaining it (motivation) at an appropriate level is challenging. The rate of non-adherence to therapy recommendations can reach 70%, especially in the case of home exercises without direct supervision by a physiotherapist [14]. A motivated patient, on the other hand, experiences less resistance to the rehabilitation programme [51]. The flexibility of therapy location and time, guaranteed by the tele system, minimises the burden imposed by traditional inpatient sessions. As a result, teletherapy participants acknowledge that teletherapy improvement provides the discipline necessary for regular exercise and contributes to high adherence among patients. In addition, they perceive benefits relating to both physical (improved performance, function, sleep) and psychological (increased confidence, motivation to achieve goals) status, which amounts to increased quality of life and reduced risk of secondary stroke [23]. Patients also often experience shrinkage of social networks after strokes. This is mainly due to loss of contact with friends, decreased participation in group events or avoidance of social activities [2,16]. This phenomenon can contribute to worsening disability, as social isolation worsens indicators of functional status and increases the risk of depression [6]. The interaction with the therapist, despite taking place in the form of video conversation, on the one hand ensures that the exercises are performed correctly and that the results are analysed, and, on the other hand, it improves the patient's mood and makes him or her feel less socially isolated. The Podury's study shows that Geriatric Depression Score significantly decreased over 12-weeks supervised home-based telerehabilitation from 3 (1–5) to 1 (0–4) ($p = 0.05$) and it also found that patients undergoing telerehabilitation had larger and more open social networks than those in the control group [2,21,22]. A correlation was observed between the size and density of the network and the improvement in patients' gait time. Social network density was also associated with improvements in upper limb motor skills, while network size was associated with reductions in depressive symptoms. Furthermore, telerehabilitation using virtual reality (VR) is characterised by a large variety of goal-oriented motor tasks due to the availability of a wide range of interactive games with progressive levels of difficulty. The motor experiences offered by VR make traditional exercise unattractive to patients. In the Sheehy study, the tele technologies were proven to be easy to use and affordable for the stroke population eligible for this type of therapy [3]. Study participants came from a wide age range and had varying levels of disability or familiarity with technology, which was not a barrier to successful participation in telerehabilitation sessions. The Chen research shows that teletherapy can be used as an efficient and user-friendly tool to deliver home-based stroke rehabilitation that enhance patients' physical recovery and mental and social-emotional wellbeing. Participants mostly reported positive experiences with the tele system. Benefits included observed improvements in limb functions, cognitive abilities, and emotional well-being. They also perceived the system to be easy to use due to the engaging experience and the convenience of conducting sessions at home [15–20,33,52,53]. Only in the publications by Cramer and Galloway were there cases of dropouts during research (Cramer—10 patients, Galloway—3 patients). The reason for this, however, was not difficulties arising from the use of the tele system but personal reasons or the need to return to work during the rehabilitation program [14,23]. Teletherapy has also had a tremendous impact on increasing the level of knowledge of strokes and secondary prevention, which are often inadequate among post-stroke patients. Through educational programmes that complement the teletherapeutic process, an optimisation of health is achieved, resulting in improved functional status and reducing the risk of secondary stroke. Thus, patient education is one of the key components of effective telerehabilitation [14]. Unfortunately, the strength of the evidence itself supporting the effectiveness of teletherapy still ranks low. The reason for this is the limited number of randomised controlled trials, small number of study participants, or heterogeneous trials [12]. Participants in the tele-improvement programme are primarily patients with mild to moderate motor disabilities who qualify for therapy based solely on specifically defined criteria (Table 2). Teletherapy itself obligatorily has to be preceded by understandable instructions and must be constantly controlled and modified by a physiotherapist. To ensure safe exercise progress, the entire improvement programme takes place in the presence of a supervisor. Nevertheless, based on the above-mentioned literature review and considering the lack of sufficient RCTs, it can be concluded that telerehabilitation is an alternative modality to conventional therapy, facilitating post-stroke patients' achievement of functional recovery and an improvement in their quality of life in the resource-limited settings caused by the SARS-CoV-2 pandemic.

Figure 2. Advantages of telerehabilitation.

Author Contributions: Concept and design of the manuscript P.M.O., collection of literature P.M.O., M.Ś., original draft preparation P.M.O., R.S.; project administration R.H.-K.; review and editing R.S., R.H.-K.; editing and final approval P.M.O., R.H.-K. All authors read and approved the final manuscript.

Funding: This research received no external funding.

Institutional Review Board Statement: Not applicable.

Informed Consent Statement: Not applicable.

Data Availability Statement: The data presented in this study are openly available in PubMed, Directory of Open Access Journals and Science Direct databases.

Conflicts of Interest: The authors declare no conflict of interest.

Abbreviations

TR	Telerehabilitation
VR	Virtual Reality
BBS	Berg Balance Scale
TUG	Timed Up-and-Go
FTSST	5 Times Sit-to-Stand
CB&M	Community Balance and Mobility Scale
SIS	Stroke Impact Scale
FMA (UE)	Fugl-Meyer Assessment Upper Extremity
FMA (LE)	Fugl-Meyer Assessment Lower Extremity

References

1. Varghese, C.; Onuma, O.; Johnson, W.; Brainin, M.; Hacke, W.; Norrving, B. Organizational update: World Health Organization. *Stroke* **2017**, *48*, e341–e342. [CrossRef] [PubMed]
2. Podury, A.; Raefsky, S.M.; Dodakian, L.; McCafferty, L.; Le, V.; McKenzie, A.; See, J.; Zhou, R.J.; Nguyen, T.; Vanderschelden, B.; et al. Social network structure is related to functional improvement from home-based telerehabilitation after stroke. *Front. Neurol.* **2021**, *12*. [CrossRef]
3. Sheehy, L.; Taillon-Hobson, A.; Sveistrup, H.; Bilodeau, M.; Fergusson, D.; Levac, D.; Finestone, H. Home-based virtual reality training after discharge from hospital-based stroke rehabilitation: A parallel randomized feasibility trial. *Trials* **2019**, *20*, 1–9. [CrossRef] [PubMed]
4. Berg-Weger, M.; Morley, J.E. Loneliness and social isolation in older adults during the COVID-19 pandemic: Implications for gerontological social work. *J Nutr. Health Aging* **2020**, *14*, 1–3. [CrossRef]
5. Negrini, S.; Grabljevec, K.; Boldrini, P.; Boldrini, C.; Moslavac, S.; Zampolini, M.; Christodoulou, N. Up to 2.2 million people experiencing disability suffer collateral damage each day of Covid-19 lockdown in Europe. *Eur. J. Phys. Rehabil. Med.* **2020**, *56*, 361–365.
6. Dhand, A.; Luke, D.; Lang, C.; Tsiaklides, M.; Feske, S.; Lee, J.M. Social networks and risk of delayed hospital arrival after acute stroke. *Nat. Comm.* **2019**, *10*, 1206. [CrossRef]
7. Markus, H.; Brainin, M. COVID-19 and stroke–a global world stroke organization perspective. *Int. J. Stroke* **2020**, *15*, 361–364. [CrossRef]
8. Tenforde, S.A.; Zafonte, R.; Hefner, J.; Iaccarino, M.A.; Silver, J.; Paganoni, S. Efficacy of home-based telerehabilitation versus in-clinic therapy for adults after stroke. *Am. J. Phys. Med. Rehabil.* **2020**, *99*, 764–765. [CrossRef]
9. Laver, K.E.; Adey-Wakeling, Z.; Crotty, M.; Lannin, N.A.; George, S.; Sherrington, C. Telerehabilitation services for stroke. *Cochrane Database Syst. Rev.* **2020**, *1*, CD010255. [CrossRef]
10. Laver, K.E.; George, S.; Thomas, S.; Deutsch, J.E.; Crotty, M. Virtual reality for stroke rehabilitation. *Cochrane DB Syst. Rev.* **2017**, *11*, CD008349. [CrossRef]
11. Moher, D.; Liberati, A.; Tetzlaff, J.; Altman, D.G. PRISMA Group; Preferred reporting items for systematic reviews and meta-analyses: The PRISMA statement. *PLoS Med.* **2009**, 6. [CrossRef]
12. Amorim, P.; Santos, B.S.; Dias, P.; Silva, S.; Martins, H. Serious Games for Stroke Telerehabilitation of Upper Limb-A Review for Future Research. *Int. J. Telerehabilitation* **2020**, *12*, 65–76. [CrossRef]
13. Burgos, P.I.; Lara, O.; Lavado, A.; Rojas-Sepúlveda, I.; Delgado, C.; Bravo, E.; Kamisato, C.; Torres, J.; Castañeda, V.; Cerda, M. Exergames and telerehabilitation on smartphones to improve balance in stroke patients. *Brain Sci.* **2020**, *10*, 773. [CrossRef] [PubMed]
14. Cramer, S.C.; Dodakian, L.; Le, V.; See, J.; Augsburger, R.; McKenzie, A.; Zhou, R.J.; Chiu, N.L.; Heckhausen, J.; Cassidy, J.M.; et al. Efficacy of home-based telerehabilitation vs in-clinic therapy for adults after stroke. A randomized clinical trial. *JAMA Neurol.* **2019**, *76*, 1079–1087. [CrossRef] [PubMed]
15. Chen, Y.; Chen, Y.; Zheng, K.; Dodakian, L.; See, J.; Zhou, R.; Chiu, N.; Augsburger, R.; McKenzie, A.; Cramer, S.C. A qualitative study on user acceptance of a home-based stroke telerehabilitation system. *Top. Stroke Rehabil.* **2020**, *27*, 81–92. [CrossRef] [PubMed]
16. Canadian Stroke Congress 2019 Abstract Supplement. *Int. J. Stroke* **2019**, *14*, 3–52. [CrossRef]
17. Dodakian, L.; McKenzie, A.L.; Le, V.; See, J.; Pearson-Fuhrhop, K.; Burke Quinlan, E.; Zhou, R.J.; Augsberger, R.; Tran, X.A.; Friedman, N.; et al. A home-based telerehabilitation program for patients with stroke. *Neurorehabil. Neural Repair* **2017**, *31*, 923–933. [CrossRef]
18. Richmond, T.; Peterson, C.; Cason, J.; Billings, M.; Terrell, E.A.; Lee, A.C.W.; Towey, M.; Parmanto, B.; Saptano, A.; Cohn, E.R.; et al. American Telemedicine Association's principles for delivering telerehabilitation services. *Int. J. Telerehabil.* **2017**, *9*, 63–68. [CrossRef]
19. Benjamin, E.J.; Blaha, M.J.; Chiuve, S.E.; Cushman, M.; Das, S.R.; Deo, R.; De Ferranti, S.D.; Floyd, J.; Fornage, M.; Gillespie, C.; et al. Heart disease and stroke statistics–2017 update: A report from the American Heart Association. *Circulation* **2017**, *135*, e146–e603. [CrossRef]

20. Dhand, A.; Lang, C.E.; Luke, D.A.; Kim, A.; Li, K.; McCafferty, L.; Mu, Y.; Rosner, B.; Feske, S.K.; Lee, J.-M. Social network mapping and functional recovery within 6 months of ischemic stroke. *Neurorehabil. Neural Repair* **2019**, *33*, 922–932. [CrossRef]
21. Prust, M. Head-to-head comparison of social network assessments in stroke survivors. *Neurohospitalist* **2020**, *11*, 18–24. [CrossRef] [PubMed]
22. Zachrison, K.S.; Dhand, A.; Schwamm, L.H.; Onnela, J.-P. A network approach to stroke systems of care. *Circ. Cardiovasc. Qual. Outcomes* **2019**, *12*. [CrossRef] [PubMed]
23. Galloway, M.; Marsden, D.L.; Callister, R.; Nilsson, M.; Erickson, K.I.; English, C. The feasibility of a telehealth exercise program aimed at increasing cardiorespiratory fitness for people after stroke. *Int. J. Telerehabilitation* **2019**, *11*, 9–28. [CrossRef]
24. Barengo, N.C.; Antikainen, R.; Borodulin, K.; Harald, K.; Jousilahti, P. Leisure-time physical activity reduces total and cardiovascular mortality and cardiovascular disease incidence in olderz adults. *J. Am. Geriatr. Soc.* **2017**, *65*, 504–510. [CrossRef] [PubMed]
25. Prout, E.C.; Mansfield, A.; McIlroy, W.E.; Brooks, D. Patients' perspectives on aerobic exercise early after *stroke. Disabil. Rehabil.* **2017**, *39*, 684–690. [CrossRef]
26. Ragupathi, L.; Stribling, J.; Yakunina, Y.; Fuster, V.; McLaughlin, M.A.; Vedanthan, R. Availability, use, and barriers to cardiac rehabilitation in LMIC. *Global Heart* **2017**, *12*, 323–334. [CrossRef] [PubMed]
27. Reinholdsson, M.; Palstam, A.; Sunnerhagen, K.S. Prestroke physical activity could influence acute stroke severity (part of PAPSIGOT). *Neurology* **2018**, *91*. [CrossRef]
28. Galloway, M. An Exploration of Low Doses of Exercise on Cardiorespiratory Fitness in People with Chronic Stroke. Ph.D. Thesis, University of Newcastle, Callaghan, Australia, 2019. (Unpublished).
29. Jackson, S.; Mercer, C.; Singer, B.J. An exploration of factors influencing physical activity levels amongst a cohort of people living in the community after stroke in the south of England. *Disabil. Rehabil.* **2018**, *40*, 414–424. [CrossRef]
30. Maddison, R.; Rawstorn, J.C.; Stewart, R.A.H.; Benatar, J.; Whittaker, R.; Rolleston, A.; Jiang, Y.; Gao, L.; Moodie, M.; Warren, W.; et al. Effects and costs of real-time cardiac telerehabilitation: Randomised controlled non-inferiority trial. *Heart* **2018**, *105*, 122–129. [CrossRef]
31. Tyagi, S.; Lim, D.; Ho, W.; Koh, Y.; Cai, V.; Koh, G.; Legido-Quigley, H. Acceptance of telerehabilitation by stroke patients: Perceived barriers and facilitators. *Arch. Phys. Med. Rehabil.* **2018**, *99*, 2472–2477. [CrossRef] [PubMed]
32. Zhishui, W.; Xu, J.; Yue, C.; Li, Y.; Liang, Y. Collaborative care model based telerehabilitation exercise training program for acute stroke patients in China: A randomized controlled trial. *J. Stroke Cerebrovasc. Dis.* **2020**, *29*. [CrossRef]
33. Chen, J.; Jin, W.; Dong, W.S.; Jin, Y.; Qiao, F.L.; Zhou, Y.F.; Ren, C.C. Effects of home-based telecommunication rehabilitation on physical function for stroke survivors with hemiplegia: A random controlled trial. *Am. J. phys. Med. Rehabil.* **2017**, *96*, 152–160. [CrossRef]
34. Torrisi, M.; Maresca, G.; Cannavò, D.A.; Sciarrone, F.; Gemelli, P.; Silvestri, G.; Bramanti, A.; De Luca, R.; Calabrò, R.S. Using telerehabilitation to improve cognitive function in post-stroke survivors: Is this the time for continuity of care? *Int. J. Rehabil. Res.* **2019**, *42*, 344–351. [CrossRef] [PubMed]
35. De Luca, R.; Leonardi, S.; Spadaro, L.; Russo, M.; Aragona, B.; Torrisi, M.; Maggio, M.G.; BioEng, A.B.; Naro, A.; De Cola MStat, M.C.; et al. Improving cognitive function in patients with stroke: Can computerized training be the future? *J. Stroke Cerebrovasc. Dis.* **2018**, *27*, 1055–1060. [CrossRef]
36. Venkatesh, V.; Thong, J.Y.; Xu, X. Davis User Acceptance of Information Technology: Toward a Unified View. *MIS Q.* **2003**, *27*, 425. [CrossRef]
37. Noh, H.-J.; Lee, S.H.; Bang, D.H. Three-dimensional balance training using visual feedback on balance and walking ability in subacute stroke patients: A single-blinded randomized controlled pilot trial. *J. Stroke Cerebrovasc. Dis.* **2019**, *28*, 994–1000. [CrossRef]
38. Yang, C.M.; Wang, Y.-C.; Lee, C.-H.; Chen, M.-H.; Hsieh, C.-L. A comparison of test–retest reliability and random measurement error of the Barthel Index and modified Barthel Index in patients with chronic stroke. *Disabil. Rehabil.* **2020**, 1–5. [CrossRef]
39. Rose, D.K.; DeMark, L.; Fox, E.J.; Clark, D.J.; Wludyka, P. A backward walking training program to improve balance and mobility in acute stroke. *J. Neurol. Phys. Ther.* **2018**, *42*, 12–21. [CrossRef] [PubMed]
40. Sarfo, F.S.; Ulasavets, U.; Opare-Sem, O.K.; Ovbiagele, B. Telerehabilitation after stroke: An updated systematic review of the literature. *J. Stroke Cereb. Dis.* **2018**, *27*, 2306–2318. [CrossRef] [PubMed]
41. Tchero, H.; Tabue, M.; Lannuzel, A.; Rusch, E. Telerehabilitation for stroke survivors: Systematic review and meta-analysis. *J. Med. Internet. Res.* **2018**, *20*. [CrossRef] [PubMed]
42. Simpson, B.D.; Bird, M.L.; English, C.; Gall, S.L.; Breslin, M.; Smith, S.; Schmidt, M.; Callisaya, M.L. Connecting patients and therapists remotely using technology is feasible and facilitates exercise adherence after stroke. *Top. stroke Rehabil.* **2020**, *27*, 93–102. [CrossRef] [PubMed]
43. Sousa de, D.G.; Harvey, L.A.; Dorsch, S.; Glinsky, J.V. Interventions involving repetitive practice improve strength after stroke: A systematic review. *J. Physiother.* **2018**, *64*, 210–221. [CrossRef]
44. Ameer, K.; Ali, K. iPad use in stroke neuro-rehabilitation. *Geriatrics* **2017**, *2*, 2. [CrossRef] [PubMed]
45. Miller, K.K.; Porter, R.E.; DeBaun-Sprague, E.; Van Puymbroeck, M.; Schmid, A.A. Exercise after stroke: Patient adherence and beliefs after discharge from rehabilitation. *Top. Stroke Rehabil.* **2017**, *24*, 142–148. [CrossRef] [PubMed]

46. Odetunde, O.M.; Binuyo, O.T.; Maruf, F.A.; Ayenowowon, S.O.; Okonji, A.M.; Odetund, N.A.; Mbada, C.E. Development and feasibility testing of video home based telerehabilitation for stroke survivors in resource limited settings. *Int. J. Telerehabilitation* **2020**, *12*, 125–135. [CrossRef]
47. Adhikari, S.P.; Shrestha, P.; Dev, R. Feasibility and effectiveness of telephone-based telephysiotherapy for treatment of pain in low-resource setting: A retrospective prepost design. *Pain Res. Manag.* **2020**, *2020*. [CrossRef]
48. Donkor, E.S. Stroke in the21stCentury: A Snapshot of the Burden, Epidemiology, and Quality of Life. *Stroke Res. Treat.* **2018**, *2018*, 1–10. [CrossRef]
49. Sarfo, F.S.; Ulasavets, U.; Opare-Sem, O.K.; Ovbiagele, B. Pilot trial of a tele-rehab intervention to improve outcomes after stroke in Ghana: A feasibility and user satisfaction study. *J. Neurol. Sci.* **2018**, *387*, 94–97. [CrossRef]
50. Zhao, J.; Rudd, A.; Liu, R. Challenges and potential solutions of stroke care during the coronavirus disease 19 (COVID-19) pandemic. *Stroke* **2020**, *51*, 1356–1357. [CrossRef]
51. Jeffers, M.S.; Karthikeyan, S.; Gomez-Smith, M.; Gasinzigwa, S.; Achenbach, J.; Feiten, A. Does stroke rehabilitation really matter? part b: An algorithm for prescribing an effective intensity of rehabilitation. *Neurorehabil. Neural Repair* **2018**, *32*, 73–83. [CrossRef] [PubMed]
52. Chen, J.; Sun, D.; Zhang, S.; Shi, Y.; Qiao, F.; Zhou, Y.; Liu, J.; Ren, C. The effects of home-based telerehabilitation in patients with stroke: A randomized controlled trial. *Neurology* **2020**, *95*, e2318–e2330. [CrossRef] [PubMed]
53. Chen, Y.; Abel, K.T.; Janecek, J.T.; Chen, Y.; Zheng, K.; Cramer, S.C. Home-based technologies for stroke rehabilitation: A systematic review. *Int. J. Med. Inform.* **2019**, *123*, 11–22. [CrossRef] [PubMed]

MDPI
St. Alban-Anlage 66
4052 Basel
Switzerland
Tel. +41 61 683 77 34
Fax +41 61 302 89 18
www.mdpi.com

Healthcare Editorial Office
E-mail: healthcare@mdpi.com
www.mdpi.com/journal/healthcare